Handbook of Petroleum Refining Processes

Editor

Naitik Verma

scitus academics

Handbook of Petroleum Refining Processes
Edited by **Naitik Verma**

Printed in 2017

ISBN: 978-1-68117-396-2

Library of Congress Control Number: 2015941586

SCITUS Academics LLC,
616, Corporate Way, Suite 2, 4766,
Valley Cottage, NY 10989

www.scitusacademics.com

Handbook of Petroleum Refining Processes

Contents

Preface

Fundamentals of Petroleum Refining presents the fundamentals of thermodynamics and kinetics, and it explains the scientific background essential for understanding refinery operations. The text also provides a detailed introduction to refinery engineering topics, ranging from the basic principles and unit operations to overall refinery economics. The book covers important topics, such as clean fuels, gasification, biofuels, and environmental impact of refining. Petroleum refining is a unique and critical in the petroleum supply chain, from the wellhead to the pump. The other add value to petroleum mainly by moving and storing it (e.g., lifting crude oil to the surface; moving crude oil from oil fields to storage facilities and then to refineries; moving refined products from refinery to terminals and end-use locations, etc.). Refining adds value by converting crude oil (which in itself has little end-use value) into a range of refined products, including transportation fuels. The primary economic objective in refining is to maximize the value added in converting crude oil into finished products.

Editor

Chapter 1

The Design of Water-using Systems in Petroleum Refining Using a Water-pinch Decomposition

Alberto Alva-Argáez[a], Antonis C. Kokossis[b], and Robin Smith[c]

[a]CETC-Varennes, Natural Resources Canada, 1615 Lionel Boulet Blvd., Varennes Que., Canada

[b]Process & Information Systems Engineering, School of Engineering, University of Surrey, Guildford, Surrey, UK

[c]The University of Manchester, Manchester, UK

ABSTRACT

Water reuse and recycling offer substantial potential for savings in petroleum refining, as the water volumes processed are large.

Presently, there is a lack of methods to systematically screen and analyze design alternatives using a total systems approach. Such an approach would consider effluent treatment, recycle of treated water and freshwater distribution simultaneously. The paper contributes with a systematic methodology that empowers conceptual engineering and water-pinch [R. Smith, Chemical Process Design and Integration, 2nd ed., John Wiley & Sons, 2005; Y.P. Wang, R. Smith, Wastewater minimization, Chem. Eng. Sci. 49 (7) (1994) 981–1006; Y.P. Wang, R. Smith, Design of distributed effluent treatment systems, Chem. Eng. Sci. 49 (18) (1994) 3127–3145.] with mathematical programming methods. The method focuses on petroleum refineries explaining trade-offs and savings between freshwater costs, wastewater treatment, piping costs and environmental constraints on the discharge.

INTRODUCTION

The total amount of water used in refineries has been estimated to an average 65–90 gallons of water per barrel of crude oil [7]. The waste effluents from petroleum refineries typically require treatment before reuse or discharge. Stringent regulations on the discharge are likely to become stricter, with restrictions applying not only to industrial users, but also to municipal wastewater treatment operations. Industry faces a challenge to reduce the wastewater it generates and attain sustainable standards of operation. Designs should incorporate economical solutions that address effluent segregation systems and the regeneration of effluents. The concentration of wastewater pollutants depends on the amount of process steams, as well as the amount and the composition of the process and cooling water in the plant. Water reuse and recycling define the final concentrations of pollutants. The standard practice is to bring together contaminated wastewater into a single wastewater stream [7] and [37] and treat it further in a central treatment facility. The practice produces streams that are more difficult to treat and invariably leads to higher processing costs. There is an apparent lack of methods to systematically screen and analyze design alternatives with a total systems approach. Such an approach would consider effluent treatment, recycle of treated water and freshwater distribution simultaneously. The concept of mass exchange networks [8], [9], [10], [11],[18] and [32] should be attributed the first systematic approach to the problem. A methodology to design an optimal water network has been presented in [2]. Koppol et al. [25] investigated the impact of pursuing zero-discharge policies. MINLP formulations have been previously documented in the literature [14] to report challenges to handle the non-linear elements of the formulations. Early work to tackle bilinear terms includes applications of recursive LP and successive linear programming (SLP) [19], [20], [21], [22],[31] and [44]. Successful applications have been published in [3]. More systematic approaches to handle non-linearities are found in algorithms that employ mathematical decompositions based on generic schemes [1], [4], [12], [13], [15], [36] and [39]. Examples include Branch and Bound strategies, cutting plane methods and, more recently genetic algorithms applied to small problems [35].

Water-pinch is now an established concept in the literature. It is a targeting approach first developed by Wang and Smith [40] and [41] and later improved by a number of researchers [2], [17], [18], [26], [27], [28],[29] and [38] for applications in water re-use and wastewater minimisation. Water-pinch introduced the limiting water profile in graphical representations based upon concentration and mass loads. For each water consumer, water-pinch uses the maximum allowable inlet concentration, maximum allowable outlet concentrations and the mass load of each pollutant to be removed (or the water flow rate). A limiting composite curve can be constructed by combining all individual profiles into a single composite curve. Matching a water supply line provides targets for minimum freshwater consumption ahead of design. A manual design procedure has been developed to achieve targets.

A novel decomposition approach is advocated here that simplifies the challenges of the optimization problem, making systematic use of water-pinch insights to define successive projections in the solution space. The work builds on the propositions by Alva-Argaez et al. [2] and Gunaratnam et al. [17]. The approach employs a strategy to address mixed-integer linear programming formulations where binary variables enable the handling of network complexity and the imposition of practical constraints to ensure meaningful solutions. The formulation essentially extends the capabilities of the already powerful water-pinch analysis into multiple contaminants, accounting for capital costs in the objective, and the study of trade-offs between freshwater usage, mass exchanger costs, and the pipe network installation. The optimization directly accounts for cost elements and overcomes inherent limitations in multiple contaminants. The paper concludes with a generic methodology for the water system design of petroleum refineries and an approach with a potential to address retrofit problems.

WATER USERS IN A PETROLEUM REFINERY

Refineries employ a wide spectrum of solvents (and contact solvent processes) with differential solubilities to extract desirable and undesirable feedstock components. The processes produce wastewater and are invariably designed to optimize the use, re-use and the recycles associated with each individual solvent. Oil and solvents are the main pollutants in the operation whereas the bottoms of the fractionation towers represent the major source of wastewater. Water users in refineries are widely reviewed in the literature [5],[7], [11], [16], [30], [33], [34], [37] and [42] and summarized for the purposes of this study in Table 1.

Table 1: Major water users in petroleum refining

Water users	Water source	Wastewater main pollutants	Comments
Crude desalting	Fresh or foul water stripped	Free oil, ammonia, sulfides and suspended solids	3–10 vol.% on crude charge

Crude oil distillation	Stripping steam	Sulfides, ammonia, phenols, oil, chlorides, mercaptans	Oil sampling lines produce significant flows
Thermal cracking	Overhead accumulators	H_2S, ammonia, phenols	
Catalytic cracking	Stripping steam	Oil, sulfides, phenols, cyanide, ammonia	Alkaline wastewaters
Hydrocracking	Stripping steam	High in sulfides	
Polymerization	Pretreatment for H_2S removal	Sulfides, mercaptans, ammonia	Low volume but high strength waste
Alkylation	Caustic/water wash	Spent caustic, oil, sulfides	
Isomerization	Low demand	Low level of phenols	No sulfides, ammonia present
Reforming	Low demand	sulfide	Relatively clean process
Hydrotreating	Stripping steam	Ammonia, sulfides, phenol	Effluent depends heavily on feedstock

In crude desalting the crude oil is dehydrated using a combination of emulsion-breaking and coalescence. The amount of water present in the crude oil is approximately 0.1–2 vol.% and the salts contained in the emulsified aqueous phase range from 10 to 250 pounds per thousand barrels (p.t.b). Salts are separated from oil either in the presence of chemicals (followed by heating and gravity separation) or under the influence of high voltage electrostatic fields that agglomerate the dispersed droplets. A single-stage electrical desalter can reduce the salt content down to 3 p.t.b. A two-stage unit can reduce it down to 0.3 p.t.b. Dehydration down to about 0.2 vol.% water on crude can usually be achieved. Make-up wash water is always added to the crude oil to assist the desalting process and it is used in the range of 3–10 vol.% on crude.

In crude oil distillation the wastewater comes from: (i) overhead accumulators (prior to recirculation or transfer from hydrocarbons to other fractionators) containing sulfides, ammonia, phenols, oil, chlorides and mercaptans; (ii) oil sampling lines; (iii) stable emulsions from barometric condensers used to create vacuum conditions (as barometric condensers are replaced by surface condensers, the oil vapours are not in contact with water); (iv) the overhead reflux drum where stripping steam is condensed with naphtha and contains H_2S and NH_3 (mostly as NH_4SH); (v) the overhead product drum that contains H_2S and possibly phenols. In thermal cracking, the major source of wastewater is the overhead accumulator on the fractionator, where water is separated from the hydrocarbon vapour and sent to the sewer system. Effluent streams can be expected to contain H_2S, NH_3 and phenols. Catalytic cracking units, are some of the largest sources of sour and phenolic wastewaters in a refinery. Pollutants from catalytic cracking generally come from the steam strippers and overhead accumulators on fractionators used to recover and separate the various hydrocarbon fractions produced in the cracking process. The major pollutants are oil, sulfides, phenols, cyanides, and ammonia. These pollutants produce an alkaline wastewater with high BOD and COD levels.

Hydrocracking is basically catalytic cracking in the presence of hydrogen and at least one wastewater stream from the process should be high in sulfides, since hydrocraking reduces the sulfur content of the material being cracked. Most of the sulfides are in the gas products, which are sent to a treating unit for removal and/or recovery of sulfur and ammonia. In polymerization, even though the process makes use of an acid catalyst, the waste is alkaline because the catalyst is recycled and any remaining acid is removed by caustic washing. Most of the waste comes from the pretreatment stage. The wastewater is high in sulfides, mercaptans and ammonia. Catalytic polymerisation plants may require disposal facilities for spent caustic containing sodium phosphate. In alkylation the major discharge is the spent caustics from the neutralisation stage. These wastewaters contain suspended and dissolved solids, sulfides, oils and other contaminants. Water drawn off from the overhead accumulators contains varying amounts of oil, sulfides and other contaminants but it is not a major source of waste.

Isomerization wastewaters present no major pollutant discharge problems. Sulfides and ammonia are not likely to be present in the effluent. Low levels of phenols and oxygen demand also should be expected. Reforming is a relatively clean process. The volume of wastewater flow is small and none of the waste streams has high concentrations of significant pollutants. The wastewater is alkaline and the major pollutant is sulfide from the overhead accumulator on the stripping tower used to remove light hydrocarbon fractions from the reactor effluent. Principal hydrotreating processes used include the pretreatment of catalytic reformer feedstock, naphtha desulfurization, lube oil polishing, pretreatment of cat cracking feedstock, heavy gas oil and residual desulfurization and naphtha saturation. The strength and quantity of the wastewater depends on the process used and the feedstock. Ammonia and sulfides are the primary contaminants but phenols may also be present. The catalytic hydrotreating of diesel oil to reduce its sulfur content does not per se produce sour water. However, the subsequent steam stripping of the hydrotreated diesel to remove the free H_2S does yield a sour condensate overhead along with a small amount of by product sour naphtha.

WATER NETWORK SYNTHESIS IN A PETROLEUM REFINERY

A design method is proposed that is based on a superstructure model formulated as a mixed-integer non-linear programming (MINLP) problem where every water user and treatment unit are represented by an inlet mixer and an outlet splitter; freshwater sources are introduced as splitters and wastewater discharge points as final mixers (Fig. 1). The non-linearities in the formulation are due to bilinear terms that appear in the mass balances (superstructure mixers and splitters), the non-linear terms of the sizing equations, and the cost functions for the water-using operations. All water users and treatment units have been modelled following the concepts developed by Wang and Smith in the water-pinch methodology [40] and [41].

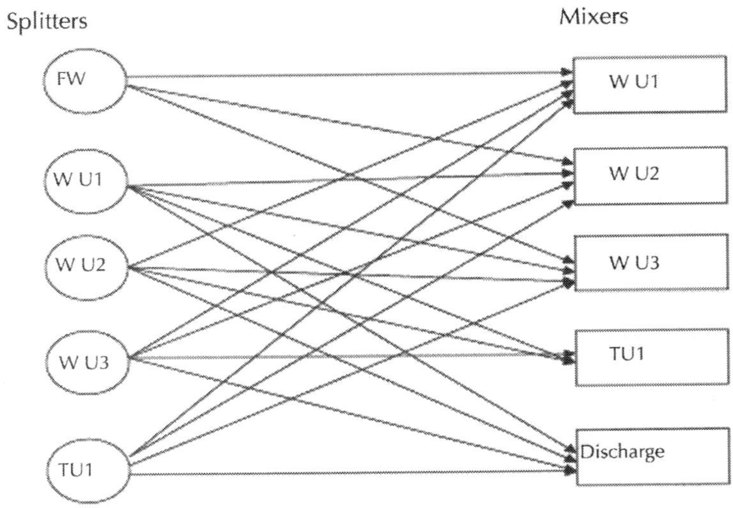

Figure 1: Schematic representation of the superstructure model (single water source FW, water users WU1, WU2 and WU3 and treatment unit TU1. A single discharge point is illustrated).

Outline of the Solution Procedure

The approach integrates water-pinch with mathematical programming. Its overall iterative solution procedure is illustrated in Fig. 2. Let $(P_1)^k$ and $(P_2)^k$ be the formulation of the superstructure model with all bilinear terms projected onto the concentration space (MILP) and the set of mass balance equations projected onto the flowrate space (LP), respectively at iteration k, the algorithm involves the following steps:

- Set k = 1. Solve problem $(P_1)^k$ with $C_{c,i}^{out,k} = C_{c,i}^{max,out,k}$ and $C_{c,i}^{out,k} = 0$ to obtain z_1^{k*} and the optimal values of all the flow rates in the network, e.g. $F_i^{eff,byp}$, F_t^{out*}, $F_{t,t'}^{it*}$ and $F_{i,t'}^{tr*}$. Let $F_i^{eff,byp,k} = F_i^{eff,byp*}$, $F_t^{out,k} = F_t^{out*}$, $F_{t,t'}^{it,k} = F_{t,t'}^{it\ *}$ and $F_{i,t}^{tr,k} = F_{i,t}^{tr*}$.

- Solve problem $(P_2)^k$ for a new vector of outlet concentrations, $C^{out,*}_{c,i}$. Set $C^{out,k+1}_{c,t} = C^{out,*}_{c,t}$.
- Set $k = k + 1$, $pk^{+1} = p^k \times 10$ (penalty value in objective function), and solve problem $(P_1)k^{+1}$ to obtain new values of the flow rates in the network. Let the optimal objective value be z_1^{k+1}.
- If $\chi^{inf \cdot k} \leq \varepsilon_2$ or $\left| z^k_{opt} - z^{k+1}_{opt} / z^k_{opt} \right| \leq \varepsilon_1$, stop. Otherwise, go to step 2.
- Solve the NLP that results from problem $(P_2)^k$ with all structural features fixed according to the solution from step 4.

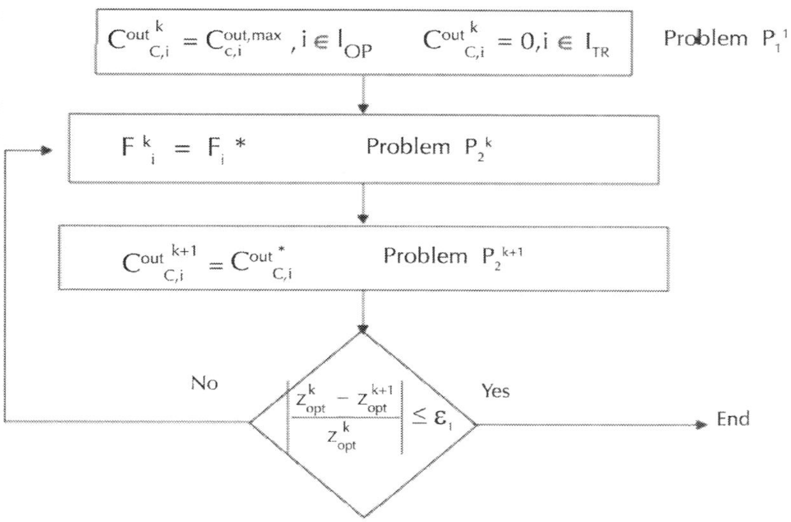

Figure 2: Iterative procedure to find a feasible solution.

The method iterates between the two problems:

- the change in the objective function of $(P_1)^k$ less than a tolerance, ε_1 (typically around 5% change), and
- feasibility is reached, i.e., the term ${}^{inf,}k$ is sufficiently small (tolerance ε_2).

The solution stands as an initial point for an NLP optimization of the un-projected solution. The implementation of the algorithm relied on the GAMS modelling environment [6], OSL was the solver selected for all MILP and LP formulations and CONOPT was employed for the NLP stage. In order to specify the vector of outlet concentrations in all water users, the approach builds on the insights from water-pinch methodologies where all contaminants tend to reach their maximum limits and, in the optimal solution, at least one contaminant will be at its maximum limit. The initial assumption of perfect treatment performance defines an infeasible point for the original problem (P), as no treatment unit can have$RR_{c,t}$ = 1. To account for the infeasibilties, a penalty function is introduced in the objective function of (P_1), which drives the distance from the feasible region to zero. The uncoupling of the bilinear terms (flow times concentration) with the introduction of the mass flows and the relaxation-projection strategy allows the search to be performed in the space defined by the convex feasible region of problem $(P_1)^k$. The choice of weights for the ^{inf} term follows the suggestions made in [31]. In all the examples, the alternative assignment of p-values as preset multiples (usually 10^3 times) of the largest of the Lagrange multipliers associated with active constraints [39] and [44] yields premature convergence.

The approach is generic and applies for an arbitrary set of contaminants. The majority of problems involve: hydrocarbons (HC), hydrogen sulfide (H_2S) and salts (salts). Water users are usually among:

- atmospheric distillation unit (CDU);
- vacuum distillation operation (VDU);
- hydrotreating (HDS);
- crude desalting (desalter);
- cooling tower (C. tower);
- boiler house;
- delayed coker;
- other operations grouped together (others).

The steam stripping operation of HDS and the desalting operation are water-using operations classified as mass exchange units. The information on the streams processed with water in the mass exchange operations is given in Table 2 (key contaminants). An alternative multi-contaminant approach requires only additional information and sizing equations in the models. For mass exchangers, the number of equilibrium stages are calculated for each contaminant ($N_{c,i}$). The exchanger is sized according to the contaminant calling for the largest number of stages using:

$$Ni \geq N_{c,i} \quad \forall c \in C$$

Table 2: Data for the process streams involved in mass exchangers (case study)

Process	Contaminant	$C^{in,PS}$ (ppm)	$C^{out,PS}$ (ppm)	Flow rate (t/h)	Mass load (g/h)
HDS	HC	–	–	54.222	3,400
	H_2S	8000	350		414,800
	Salt	–	–		4,590
Desalter	HC	–	–	263.15	800
	H_2S	–	–		200
	Salt	4000	200		1,000,000

Vapour–liquid equilibria used in Kremser type of relations [24] are derived from data available from a simulation study. For the HDS and the Desalter these are assumed to be linear. More specifically, for the HDS steam stripping (key contaminant: hydrocarbons) these are:

$$y^{*}=0.6121x$$

whereas for the desalter (key contaminant is salts) this is:

$$y^{*}=0.0018x$$

In both expressions, y* is the equilibrium concentration in the process stream and x is the corresponding concentration in the water stream (in ppm). The economic criteria are applied as shown in Table 3.

Table 3: Economic criteria

Freshwater cost	**0.3$/t**
Operating time	8600 h/year
Interest rate	15%
Time of return on investment (leading to an annualisation factor of 0.438)	3 years
Installation factor for piping cost	4

The cost functions for the stripping column are given by:

$$\text{cost}^{\text{shell}}(\$) = 10,240H$$

$$\text{cost}^{\text{tray}}(\$/\text{tray}) = 490e^{0.8D}$$

H and D indicate the column height and diameter (in m). The cost of the trays is approximated by a piecewise linear expression of the form:

$$D = \sum_q D_q^F \lambda_q$$

$$\text{cost}^{\text{tray}} = \sum_q \text{cost}_q^F \lambda_q \quad \text{for } q = 1, 2, \ldots, N_q$$

The set q is used to define the number of linear segments. For the column diameter, a range between 0.5 and 4 m is considered (i.e. $D_1^F = 0.5$ and $D_{N_q}^F = 4$) with two linear segments (i.e., Nq = 3). The number of real trays is calculated using a 60% overall column efficiency. The desalting cost is expressed by:

- A fixed capital cost for each desalting stage of $722,895 and
- An operating cost of $612,262/year per stage (to account for electricity and chemicals usage) [16].

Cost calculations further assume:

- Stainless steel construction;
- 20 min residence time;
- Horizontal cylindrical tank;
- L/D = 3.

The distances (in m) between operations are given in Table 4. Carbon steel piping is assumed as a standard and the cost of the pipework is calculated by [23]:

Costpipe=3,603.4·A+124.46 [$/m length]

Table 4: Matrix of distances for the oil refinery

Source	Sink								
	S1	O1	O2	O3	O4	O5	O6	O7	Discharge
S1	0	25	225	50	40	35	200	250	1000
O1	25	0	300	50	75	450	100	150	600
O2	225	300	0	60	100	65	225	250	500
O3	50	50	60	0	250	150	75	350	400
O4	40	75	100	250	0	350	175	100	325
O5	35	100	65	150	350	0	125	75	150
O6	200	100	125	75	175	125	0	275	125
07	250	150	250	350	100	75	275	0	50

Base Case

The operating data is given in Table 5 and the total water use is 1250 t/h. Even though the base case does not consider optimization, it still accounts for some water reuse, as the effluent from the CDU steam stripping feeds the single-stage desalter. A fixed water loss of 770 t/h is assumed for the evaporative cooling system; the remaining flow is available for reuse and treatment. The flow in the cooling system can be reduced to 813.3 t/h without mplications on the heat removal of the plant. The costs associated with the base case are presented as follows:

- Column O2: (10 trays) Total cost of $150,454, annualised cost of $65,898/year.
- Desalter O3: (single-stage) Total cost of $722,896, annualised cost of $316,629 plus annual operating cost o the unit of $612,262/year. In total, $928,891/year.

- Water cost is $3,225,000/year.
- Pipework cost (assumed flow velocity, 2 m/s everywhere) $426,412/year.

Table 5: Base case design

Operation	Flow (t/h)	Contaminant	C^{in} (ppm)	C^{out} (ppm)	Load (g/h)
CDU	50.00	HC	0.00	13.50	675.00
		H_2S	0.00	360.00	18000.00
		Salt	0.00	31.50	1575.00
HDS	48.00	HC	0.00	70.83	3400.00
		H_2S	0.00	8641.67	414800.00
		Salt	0.00	95.63	4590.00
Desalter	10.00	HC	13.50	93.63	801.28
		H_2S	360.00	380.03	200.32
		Salt	31.50	100031.50	1000000.00
Others	25.00	HC	0.00	16.72	418.00
		H_2S	0.00	91.20	2280.00
		Salt	0.00	22.80	570.00
Cooling tower	900.00	HC	0.00	10.83	9750.00
		H_2S	0.00	1.44	1300.00
		Salt	0.00	14.44	13000.00
Boiler house	190.00	HC	0.00	0.00	0.00
		H_2S	0.00	0.00	0.00
		Salt	0.00	1789.47	340000.00
Delayed coker	37.00	HC	0.00	133.24	4930.00
		H_2S	0.00	2727.57	100920.00
		Salt	0.00	156.76	5800.00

The total annual cost of the base case is $4,645,202/year. The limiting water profile is obtained through a combination of equilibrium considerations (key contaminants only) and considerations of corrosion and fouling [5], [26], [27], [28], [29] and [33]. Table 5 presents the limiting water profile data. The outlet concentrations in the base case are lower than the maximum limits identified in limiting data set, opening up opportunities to improve water efficiency.

The design of the base case is shown in Fig. 3 and uses a single-stage desalting unit. The stripping column in HDS has 10 trays for the separation.

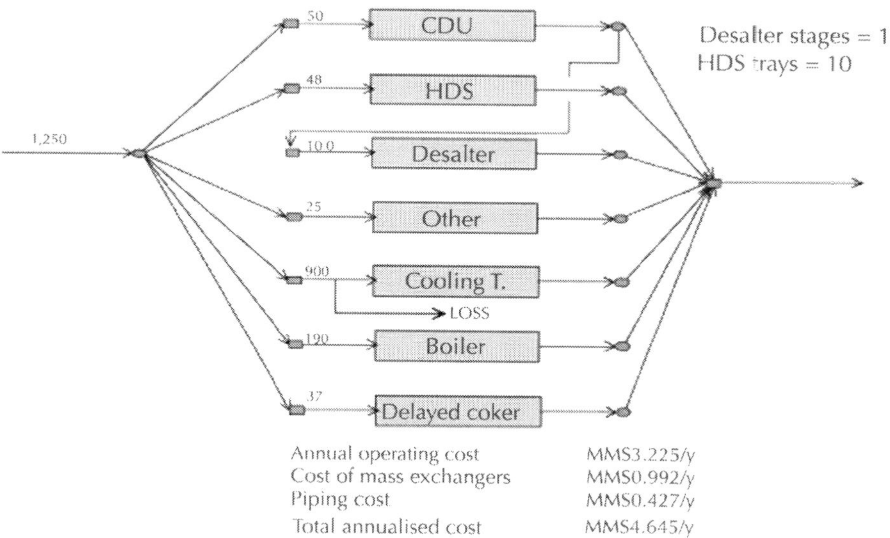

Figure 3: Base case design.

Analysis

A number of scenarios are explored to illustrate the importance of the systems approach. The variants produced consider different objectives and illustrate the potential of the approach to systematically produce optimal solutions. The problem is first solved to minimize the water consumption without and with maximum reuse (cases A and B, respectively). Next, the objective is augmented to consider capital cost associated with mass exchangers (neglecting piping and layout costs) (case C). Finally, a design is produced that addresses all design aspects (case D).

Case A

The optimal design is produced and shown in Fig. 4. The case reduces the overall freshwater consumption by enabling the water streams to reach the maximum possible concentration in each operation. Within the specifications given by Table 6, it is no longer possible to satisfy the requirements of the desalter purely with reused water from CDU. The outlet concentration of H_2S from CDU is too high to use in the desalter. The freshwater consumption is subsequently reduced via a once-through policy with no water re-use in place.

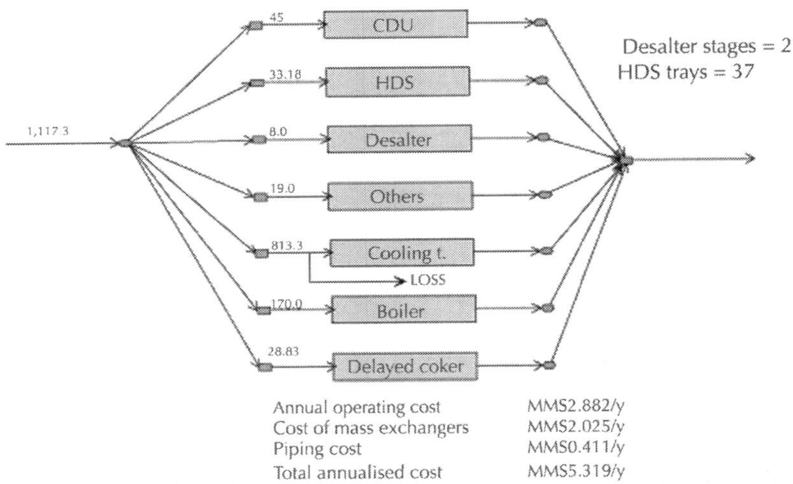

Figure 4: Design for case A.

Table 6: Limiting water profile data for the water-using operations

Process	Contaminant	$C^{in,max}$ (ppm)	$C^{out,max}$ (ppm)	Mass load (g/h)
CDU	HC	0	15	675
	H_2S	0	400	18,000
	Salt	0	35	1,575
HDS	HC	20	120	3,400

	H_2S	300	125,00	414,800
	Salt	45	180	4,590
Desalter	HC	120	220	801.2
	H_2S	20	45	200.32
	Salt	200	125,000	1,000,000
Others	HC	0	22	418
	H_2S	0	120	2,280
	Salt	0	30	570
Cooling tower	HC	150	225	9,750
	H_2S	200	310	1,300
	Salt	250	350	13,000
Boiler house	HC	0	0	800
	H_2S	0	0	200
	Salt	0	2,000	340,000
Delayed coker	HC	100	270	4,930
	H_2S	20	3,500	100,920
	Salt	50	250	5,800

The water consumption is reduced by 10%. However, the implications in the capital cost of the mass exchangers offset water savings. Even though the water is used more efficiently, the resulting design is expensive. The cost of the desalter increases significantly as it changes from a single-stage unit to a two-stage operation. It is interesting to note the operation is not as sensitive to the mass transfer driving forces (due to the high solubility of the salts in the water stream) as it is to the actual water flow rate through the unit. Thus, there is no motivation to use freshwater for desalting and increase the flow to perform the separation in a single stage is a desirable option.

Case B

The optimization problem is solved with the automated method outlined above and based on the data ofTable 5. The objective function is the minimum freshwater cost. The procedure converges in three iterations. The primal problem features 770 continuous variables and the formulation accounts for 72 binary variables. The solution consumes 54.1 CPU s overall. The NLP consumes 0.71 CPU s to converge to the solution in Fig. 5.

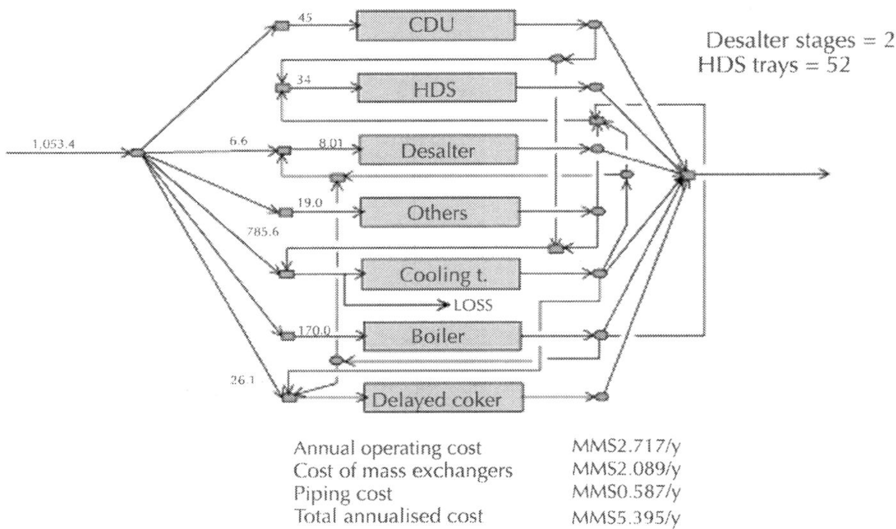

Figure 5: Design for case B.

The freshwater demand is reduced by around 15% against the base case. This is due to the maximization of the water re-use. However, the network structure is pretty complex with some of the flows being relatively small (less than 1 t/h). Constraints can be introduced to eliminate small streams as a manual manipulation of the solution. A more systematic approach is left for discussion later (case D). Alongside the freshwater reduction, an increase in the capital cost is observed and the design emerges more expensive than the once-through scenario. This is attributed to the increase in the number of stages that are selected in the mass exchangers. In the once-through case, there have been 37 trays for the HDS and two for the desalting stages. In this case the number of trays is increased to 52 for HDS and there are two stages for desalting. The increase has subsequently affected the capital cost of the stripping operation. The increased number of connections in the design should be translated into higher capital cost piping. Indeed, the latter increased by 30% as against the once-through case.

Cases A and B illustrate the fact that a design approach addressing the maximum reuse (or, equivalently, the minimization of water) may lead to rather complex designs. Although such designs can meet the targets, they are not practical or functional solutions for implementation. Combined with a general observation that the water target can be met by a multitude of nearby solutions, we step into a cost-based objective function (cases C and D) letting comparisons with the previous cases unfold from the analysis.

Case C

The objective function is formulated as the minimum total annual cost, where the cost of mass exchangers is considered. The solution is presented in Fig. 6.

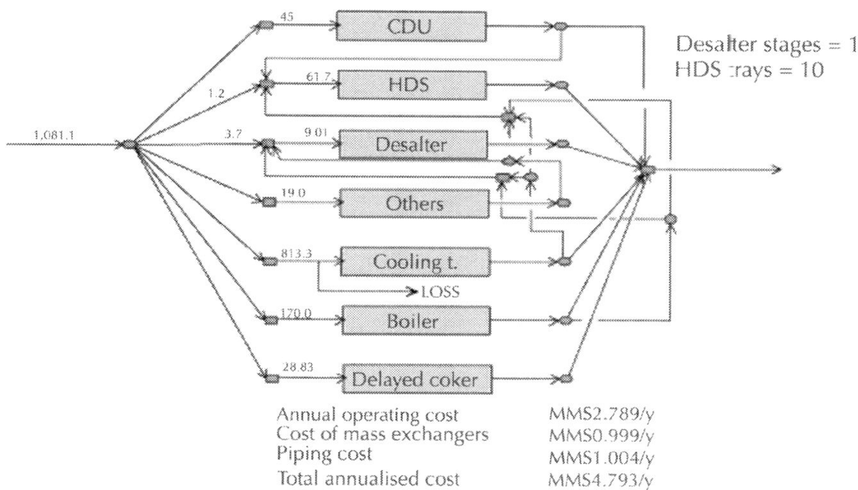

Figure 6: Design for case C.

The procedure required four iterations to converge consuming 32.1 CPU s. Primal problems featured 784 continuous and 72 binary variables. The NLP required 11.4 CPU s.

The major feature in this solution is that the cost of the mass exchangers is reduced by almost 50%. Looking at the inlets to the two mass exchange operations it can be observed that the solution provides for larger flows of water. The number of stages in HDS is reduced to ten trays (compare with 52 in the minimum freshwater cost design). The desalting operation is performed in a single stage. Since the desalting stages are major capital items, the reduction in the cost is significant, whereas the increase in the freshwater cost has been marginal (about 3% increase). The network structure still remains complex though, as characterized by the presence of small flows. Again, it is possible to eliminate the small streams as a manual manipulation of the solution.

The freshwater supply to the cooling tower is increased. The effluent though improves in quality and can be re-used into the mass exchange operations. Compare this situation with the re-use pattern obtained in the minimum freshwater cost design (Fig. 4) where the cooling tower is mainly a sink for reused water. The flow of reused water from the cooling tower to the mass exchangers is increased in the last design and achieves a lower capital cost. This design reduces freshwater as compared to the base case by 13.5%; the cost of the mass exchangers is increased by 4% as mass transfer driving forces in the steam stripping operation (HDS) are reduced. The total annual cost is 2.5% less but the piping cost has increased its share by more than 100%.

Case D

The objective is the minimum annual cost and considers explicitly freshwater, mass exchangers, and piping costs. The design obtained is shown in Fig. 7.

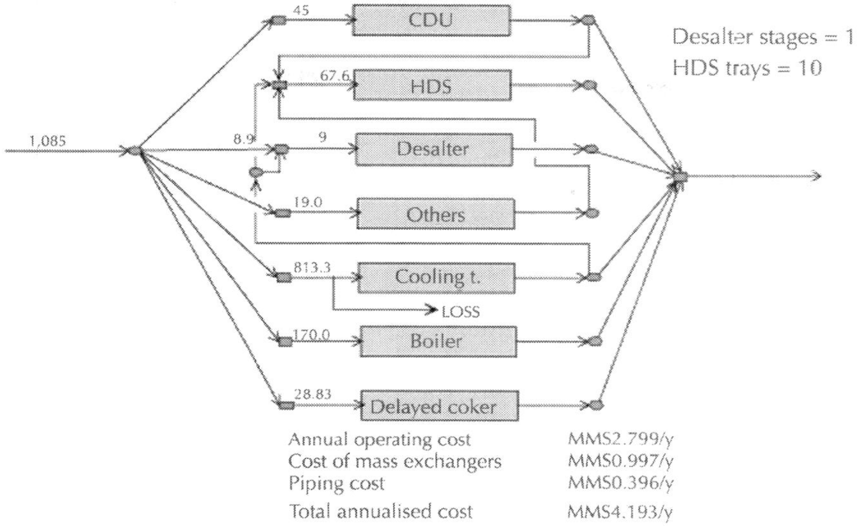

Figure 7: Design for case D.

The procedure converged in three iterations and consumed 36.1 CPU s. Primal problems comprised 792 continuous and 72 binary variables. The NLP takes 11.86 CPU s to converge. This final design shows the result when the complete picture is considered in the optimization procedure. The capital cost is similar with case C but the structure of the network is greatly simplified with the piping cost accordingly reduced. Ten trays are used in the stripping column (HDS), and a single desalting stage. The piping cost is reduced by more than 50% in comparison with case C. A summary of the results is presented in Table 7 and in Fig. 8.

Table 7: Comparison of results

Case	Operating cost (MM$/year)	Mass exchangers cost (MM$/year)	Piping cost (MM$/year)	TAC (MM$/year)
Base case	3.23	0.99	0.21	4.42
A	2.88	2.02	0.20	5.09
B	2.72	2.08	0.28	5.08

C	2.79	1.04	0.48	4.31
D	2.80	1.04	0.20	4.04

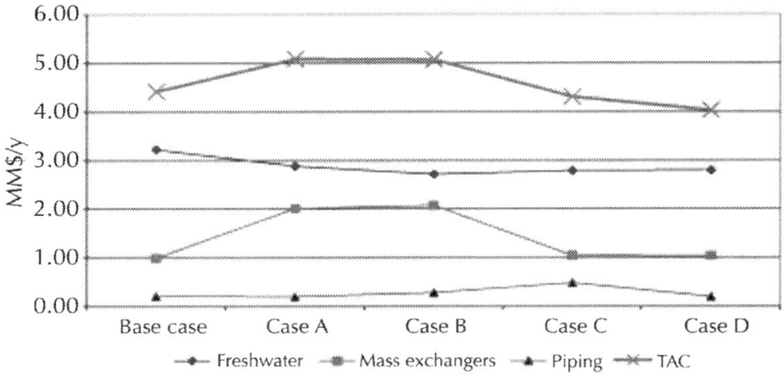

Figure 8: Comparison of results.

One should note the existence of a significant number of network configurations that are able to operate at the same water volume (major drive of the operating costs), but each featuring very different capital costs. Most notably, the size of mass exchanger units may change with the different flow patterns as a result of the different patterns for mass transfer. In the presented solutions, the desalting operation represents a major capital cost and the addition of additional stages implies a significant capital investment. In the case of the stripping column, the number of stages is less crucial.

The method can be used to explore trade-offs between operating and capital cost in the water system. As compared with the base case, the final solution achieves total annual savings of approximately $400,000 (or 8.7%). Coupled with these overall savings, freshwater consumption can be reduced by some 13%, that is 165 t/h. The calculation of the trays was found sensitive to the equilibrium parameters. The estimation of these parameters is accordingly expected to have a significant bearing in the results.

Sensitivity Analysis

A number of sensitivity experiments are presented next. The effect of an increase in the freshwater cost is illustrated in Fig. 9 where the total annualised cost (in MM$/year) is plotted as a function of the cost for the freshwater supply (in $/t).

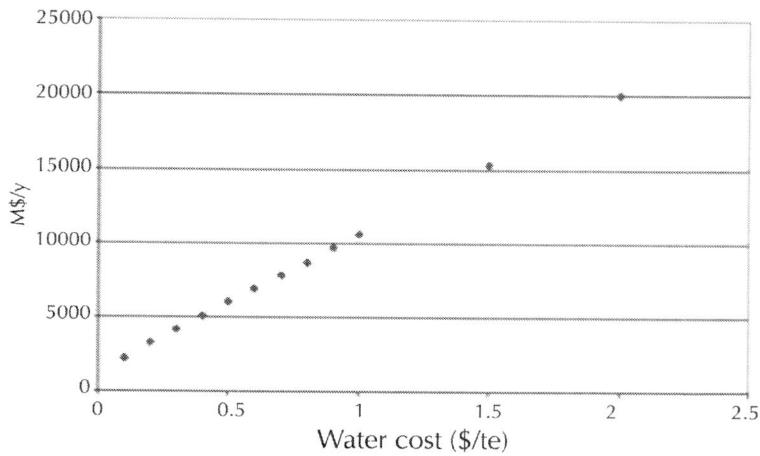

Figure 9: Sensitivity of the design to freshwater costs.

The linear relationship is explained in that the selected networks are variants of similar structures, featuring only minor changes in the flows involved. A review of freshwater intake as a function of cost demonstrates that the demanded flow remains essentially constant for a range of prices up to $2/te. Similarly, the behaviour with respect to the capital cost of the network reveals that the cost of mass exchangers features minor variations and the design appears, within the range of values analysed, robust to freshwater costs. The desalter requires a single stage. Freshwater costs should exceed $100/t before a second stage is justified. Savings of freshwater – achieved with additional stages in stripping operation – are marginal and the stripper size remains at 10 trays throughout the range of the analysis. Note that in the water network of Fig. 7

(minimum freshwater cost), the stripping operation is fed by reused water only. The stream from "Others" is consumed completely in the HDS stripper but the stream from CDU has some spare capacity.

The sensitivity of the design is finally assessed with respect to the different economic criteria employed. Designs assumed 3 years annualisation and a rate of return at 15%. Fig. 10 presents results considering different time horizons for capital return.

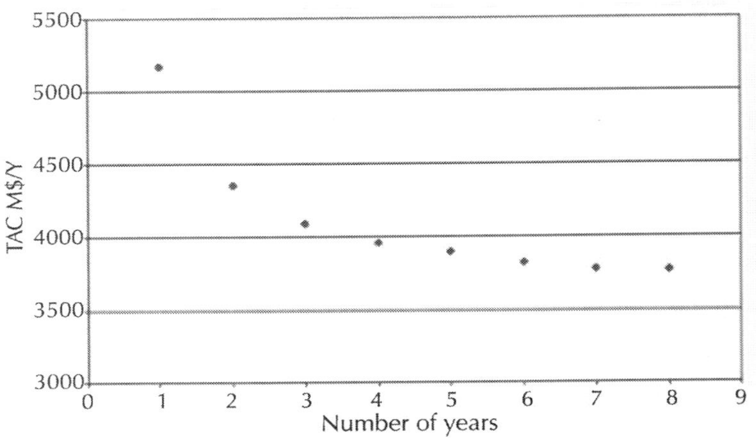

Figure 10: Total cost of the design vs. return on investment (time).

The shape of the plot explains that the water targets are robust to annualisation parameters. The freshwater consumption for the designs implied in Fig. 10 indicate designs processing similar volumes of freshwater. The underlying network structures are similar for year 1–4 and change from year 5.

The structural changes that appear in year 5 yield an alternative configuration. The designs obtained for year 4 and 5 are shown in Fig. 11 and Fig. 12, respectively. Their differences are constrained in the number of re-use connections. The design for year 4 (Design I) years features only two re-use connections (from CDU and others into HDS), whereas the design in year 5 (Design II) year shows similar connections with the addition of some reuse out of the cooling tower and into desalter and the HDS unit. Design I consumes slightly more water than Design II and is offset by savings in capital costs (mainly lower piping costs due to fewer reuse connections). Table 8 shows an overall comparison of these results.

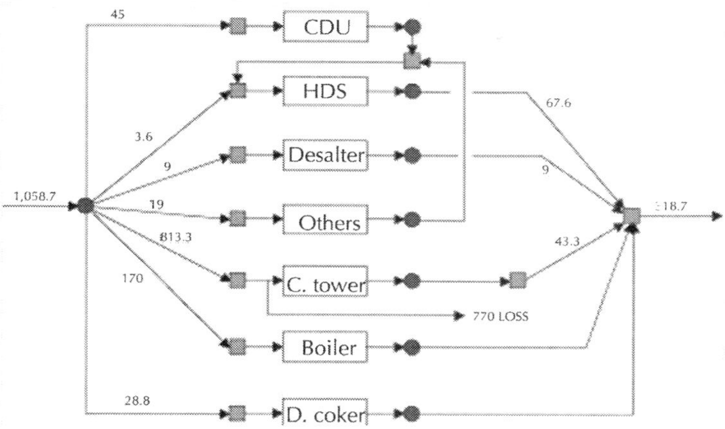

Figure 11: Four years (Design I).

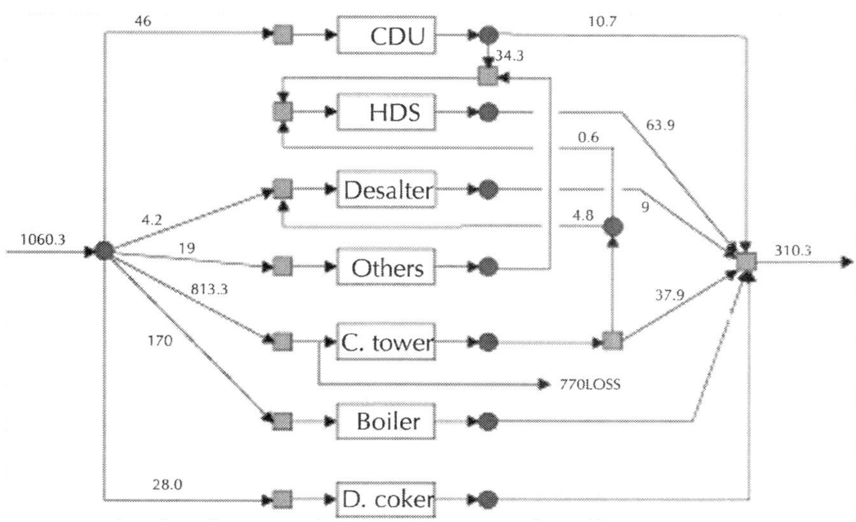

Figure 12: Five years (Design II).

Table 8: Comparison of designs

	I	II	Difference (%)
Total cost (M$/year)	4042.50	4152.82	−2.73
Freshwater cost (M$/year)	2809.02	2787.36	0.77
Piping cost (M$/year)	516.66	648.68	−25.55
ME cost (M$/year)	716.82	716.78	0.01

RETROFIT APPLICATIONS

The extension of the decomposition approach is next illustrated with retrofit applications. Most apparently, it is assumed that an existing system (units and pipe network) is already installed. In addition to the grassroots design, the model is faced with a challenge to reduce structural and operational modifications. The objective is accordingly adjusted to search for the retrofit options where the optimal use of existing capital is made.

The following modifications were required to the model.

Pipe Network

Based on the installed network, bounds are established on the flow that can potentially pass through the pipes (expressed as functions of their sizes). The existence of a pipe remains a degree of freedom and a zero cost is assigned for the use of existing pipes. Smaller or larger pipes accordingly yield penalties in the objective as they appear as "new investment" to make. In order to achieve this, a new set of constraints is introduced in the mathematical formulation. The use of special ordered sets [6] and [43] aided the modeling without a need to increase the number of binary variables. The new constraints are introduced as follows:

$$A_{i,i'}^{\mathrm{IP}} = \sum_{l} A_{i,i'}^{l} \sigma_l \qquad (1)$$

$$\text{cost}_{i,i'}^{\text{IP,pipe}} = \sum_{l} \text{cost}_{i,i'}^{\text{IP,pipe},l} \sigma_l \tag{2}$$

$$\sum_{l} \sigma_l = 1 \tag{3}$$

where the set L = {l/l is a bound for a linear segment} defines the degree of linearisation. $A_{i,i'}^{IP}$ represents the cross-sectional area of a pipe connecting operations i to i′. The $_l$(1,2, …, NL) are SOS2 variables. An alternative formulation based on binary variables is described next. Let be a 0–1 variable and let:

$$A_{i,i'}^{IPD} = A_{i,i'}^{IP} \cdot \delta_{i,i'} \tag{4}$$

$$A_{i,i'}^{IP} - A_{i,i'}^{IPUe} \delta_{i,i'} \leq 0 \tag{5}$$

$$A_{i,i'}^{IP} - UA_{i,i'}^{IP} \delta_{i,i'} \leq A_{i,i'}^{IPUe} \tag{6}$$

Namely,

$$A_{i,i'}^{IPUe} \leq A_{i,i'}^{IP} \leq UA_{i,i'}^{IP} \quad \text{then} \quad \delta_{i,i'} = 1 \quad \text{and}$$

$$\delta_{i,i'} = 0 \text{ otherwise} \tag{7}$$

$$\text{cost}_{i,i'}^{\text{IP,pipe}} = a_{i,i'}^{\text{IP,pipe}} A_{i,i'}^{IPD} + b_{i,i'}^{\text{IP,pipe}} \delta_{i,i'} \tag{8}$$

$$A_{i,i'}^{IP} - UA_{i,i'}^{IP}(1 - \delta_{i,i'}) \leq A_{i,i'}^{IPD} \tag{9}$$

$$A_{i,i'}^{IPD} \leq A_{i,i'}^{IP} \tag{10}$$

$$A_{i,i'}^{IPD} \leq UA_{i,i'}^{IP} \cdot \delta_{i,i'} \tag{11}$$

$A_{i,i'}^{IP,l}$ are the bounds on the linear segments for the cross-sectional area of the existing pipe (i,i ′). $A_{i,i'}^{IPD}$ in (4) represent bilinear terms with binary variables. Constraints (4), (5) and (6) ensure that if the cross sectional area of the pipe connecting operation i to operation i ′ exceeds the existing size, $A_{i,i'}^{IPUe}$, then $\delta_{i,i'} = 1 UA_{i,i'}^{IP}$ is the upper bound on the total flow of the connection. Once the need for a new pipe is identified, the cost function from Eq. (8) becomes linear in $A_{i,i'}^{IPD}$. Constraints (9), (10) and (11)are introduced for each existing connection.

Unit Costs

These include capital costs related to the mass exchanger units. Concerning modifications required for a staged unit (e.g. stripping column), a fixed column diameter is used as a reference (existing column) letting retrofit options consider changes in the number of trays. A solution featuring additional equilibrium stages could be translated as a recommended modification to the internals of the column (e.g. use of high efficiency trays), additional trays, or the addition of a new column. The additional capital cost is evaluated considering the cost of replacement and modifications. As in the previous cases, a piecewise linear representation that is based on SOS2 sets is employed:

$$N_i = \sum_h N_i^h \omega_h \qquad (12)$$

$$\text{cost}_i^{\text{ME}} = \sum_h \text{cost}_i^{\text{ME},h} \omega_h \qquad (13)$$

$$\sum_h \omega_h = 1 \qquad (14)$$

And alternatively, the variable β_i [0,1] is introduced to identify the number of stages in the solution. Let $N'_i = N_i \beta_i$ and:

$$N_i - (N_i^E E_i^0)\beta_i \geq 0 \qquad (15)$$

$$N_i = N_i^U \beta_i \leq (N_i^E E^0) \qquad (16)$$

$$\text{if} \quad (N_i^E E_i^0) \leq N_i \leq N_i^U, \quad \text{then,} \quad \beta_i = 1 \quad \text{and}$$
$$\beta_i = 0 \,\text{otherwise} \qquad (17)$$

$$\text{cost}_i^{\text{ME}} = N'_i C F_i \qquad (18)$$

$$N_i - N_i^U(1 - \beta_i) \leq N'_i \qquad (19)$$

$$N'_i = N_i \quad (20)$$

$$N'_i \leq N_i^U \beta_i \quad (21)$$

$$N'_i \geq 0 \quad (22)$$

Constraints (12), (13) and (14) model the existing column. The set H = {h|h is a bound for a linear segment} defines the degree of linearisation. Parameters N_i^h represent the assumed range of number of equilibrium stages; $_h$ are defined as SOS2 variables. The parameters $cost_i^{ME,h}$ in (13) correspond to cost values for the range of equilibrium stages considered. The formulation defined by constraints (15), (16), (17), (18),(19), (20), (21) and (22) involves binary variables. Parameters N_i^E represent the existing number of equilibrium stages. The lower bound is set to zero. Constraints (15) and (16) are required equilibrium stages with respect to N_i^E. The cost of the column is defined by introducing a new variable N' (18). Constraints (19), (20), (21) and (22) complete the formulation.

RETROFIT APPLICATIONS. CASE STUDY

The problem of Section 3 is selected with the existing connections listed in Table 9 (together with their maximum flows). A simple structure is assumed for the existing system with reuse connection between CDU and the Desalter. Upper bounds $A_{i,i'}^{IPUe}$, are added to the mathematical formulation for the existing connections and their cost set to zero. From the base case, the existing stripping column has 10 trays and the diameter is 2.9 m. The parameters associated with the cost of the mass exchangers in Eq. (18) are given by:

$$\mathrm{CF} = \left(\mathrm{cost}^{\mathrm{tray,HE}}\left(\frac{E^0}{E^{0,\mathrm{HE}}}\right)\right) \quad (23)$$

where E^0 is the overall column efficiency; ε_2 the tolerance value; $E^{0,HE}$the overall column efficiency using high efficiency trays; $cost^{tray,HE}$: cost of each high efficiency tray for the given diameter.

Table 9: Existing connections and maximum flows for retrofit application

Source	Sink								
	S1	O1	O2	O3	O4	O5	O6	O7	Discharge
S1	0	50	48	0	25	900	190	37	0
O1	0	0	0	10	0	0	0	0	40
O2	0	0	0	0	0	0	0	0	48
O3	0	0	0	0	0	0	0	0	10
O4	0	0	0	0	0	0	0	0	25
O5	0	0	0	0	0	0	0	0	130
O6	0	0	0	0	0	0	0	0	190
O7	0	0	0	0	0	0	0	0	37

The desalter is modelled in a simpler way (cost factor CF_{DES}) as it can operate only as a single stage (the only possibility is to add another unit). The option of not using desalter is excluded and the cost function is modified to:

$$CDES=(N-1)CFDES \tag{24}$$

The above expression yields a zero cost if a single stage is required adding penalties if a second desalting stage is selected. The retrofit problem is solved using the proposed approach and the selected design is illustrated in Fig. 13. Primal problems feature 834 continuous variables and 72 binaries. The solution algorithm requires four iterations to converge consuming 101.9 CPU s. The MINLP model features 823 continuous variables and 85 binary variables consuming 211.1 CPU s to converge. The costs associated with this design are presented in Table 10.

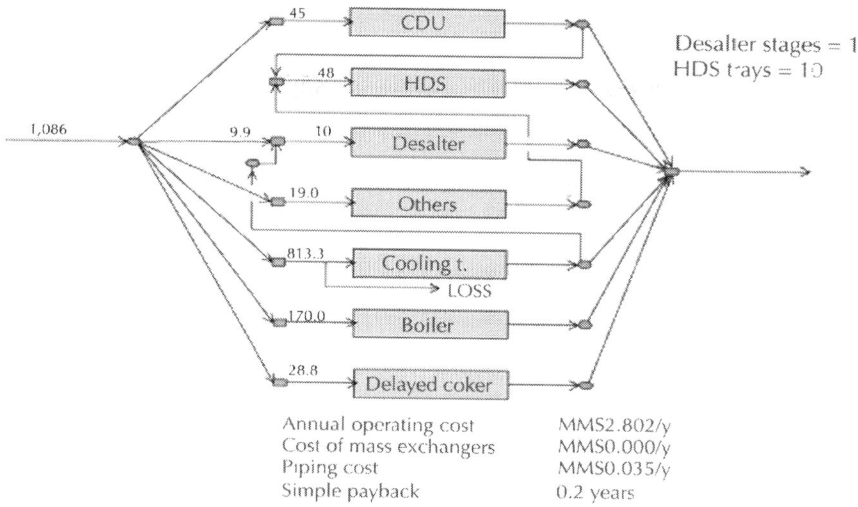

Figure 13: Retrofit design.

Table 10: Retrofit design economic data

Freshwater cost [$/year]	**2,802,086.45**
Mass exchangers [$/year]	0
Piping cost [$/year]	35,183
Total annual cost [$/year]	2,837,269.4

The freshwater consumption is reduced by 13% compared to the base case. The solution represents annual savings in freshwater costs of $423,120. The capital investment required for pipes amounts to $35,183/year. The total capital investment is $80,331 leading a payback period of 0.2 years (or approximately 3 months). According to the results presented, the pipes to install correspond to the following connections:

Source	**Sink**
S1	Desalter
CDU	HDS

Others	HDS
Cooling tower	Desalter

Existing pipes can be used for the remaining connections. The mass exchangers remain unchanged.

The method has produced results with relative ease considering the actual time spent in developing such scenarios manually for industrial projects. Solutions apparently depend upon the initial solutions but the retrofit studies carried out so far indicate that the computational burden produced is much lighter than the one accounted in the grassroots designs.

CONCLUSIONS

The paper presents a systematic approach to address water reuse in oil refineries. The methodology is based on the water-pinch decomposition and addresses a mixed-integer non-linear programming formulation that features integer variables for the connections between units. The model classifies and models major water users in refineries demonstrating its potential to address both grassroots and retrofit problems. Studies report reductions of freshwater consumption by over 10% and with minimal capital investment. The approach is further used to study the sensitivity of the solution on a variety of design and economic parameters.

The approach further enables the analysis of alternative networks useful in practical stages where several scenarios are required to screen and preview. Results indicate there is a multitude of similar solutions one needs to screen systematically. As retrofit options, the approach enables the installation of new pipes in the network, the replacement of stripping column internals with the introduction of high efficiency trays. Designs demonstrate the exhaustive search enabled by the approach. The approach is fully automated and computationally inexpensive. The design engineer can thus experiment with different scenarios, different objective functions, and can assess trade-offs between reuse performance and cost.

ACKNOWLEDGMENTS

The financial support of CONACYT, EPSRC and the Process Integration Department at UMIST is gratefully acknowledged. The support from CETC-Varennes, Natural Resources Canada for the production of the paper is also acknowledged.

REFERENCES

1. F.A. Al-Khayyal, J.E. Falk, Jointly constrained biconvex programming, Math. Oper. Res. 8 (2) (1983) 273–286.
2. Alva-Argaez, A. (1999) "Integrated design of water systems," PhD Thesis Department of Process Integration, UMIST, Manchester, UK, p. 199.
3. T.E. Baker, L.S. Lasdon, Succesive linear programming at Exxon", Manage. Sci. 31 (1985) 264–274.
4. M.S. Bazaraa, H.D. Sherali, C.M. Shetty, Nonlinear Programming: Theory and Algorithms, 2nd ed., John Wiley and Sons Inc., New York, 1993.
5. M.R. Beychok, AqueousWastes from Petroleum and Petrochemical Plants, 1st ed., John Wiley and Sons Inc., London, 1967.
6. A. Brooke, D.Kendrick, A. Meeraus,GAMS,a User's Guide, The Scientific Press, San Francisco, CA, 1988.
7. W. Byers, W. Doerr, R. Krishnan, D. Peters, How to Implement Industrial Water Reuse. A Systematic Approach, vol. 1, AIChE, New York, 1995.
8. M.M. El-Halwagi, V. Manousiouthakis, Synthesis of mass-exchange networks, AIChE. J. 8 (1989) 1233–1244.
9. M.M. El-Halwagi, V. Manousiouthakis, Automatic synthesis of massexchange networks with single component targets, Chem. Eng. Sci. 9 (1990) 2813–2831.

10. M.M. El-Halwagi, V. Manousiouthakis, Simultaneous synthesis of massexchange and regeneration networks, AIChE. J. 36 (1990) 1209–1219.
11. M.M. El-Halwagi, A.M. El-Halwagi, V. Manousiouthakis, Optimal design of dephenolization networks for petroleum refinery wastes, Trans. IChemE 70 (Part B) (1992) 131–139.
12. C.A. Floudas, A. Aggarwal, A.R. Ciric, Global optimum search for nonconvex NLP and MINLP problems, Comp. Chem. Eng. 13 (10) (1989) 1117–1132.
13. C.A. Floudas, V. Visweswaran, A global optimization algorithm (GOP) for certain classes of nonconvex NLPs-I. Theory', Comp. Chem. Eng. 14 (12) (1990) 1397–1417.
14. C.A. Floudas, Nonlinear and Mixed-Integer Optimization, 1st ed., Oxford University Press, New York, 1995.
15. B. Galan, I.E. Grossman, Optimal design of distributed wastewater treatment networks, Ind. Eng. Chem. Res. 37 (10) (1998) 4036–4048.
16. J.H. Gary, G.E. Handwerk, Petroleum Refining: Technology and Economics, 3rd ed., Marcel Dekker, New York USA, 1994.
17. M. Gunaratnam, A. Alva-Arg´aez, A.C. Kokossis, J.-K. Kim, R. Smith, Automated design of total water systems, Ind. Eng. Chem. Res. 44 (2005) 588–599.
18. N. Hallale, D.M. Fraser, Capital cost targets for mass exchange networks. A special case: water minimisation, Chem. Eng. Sci. 53 (2) (1998) 293–313.
19. C.A. Haverly, Recursion model behaviour. More studies, ACM Sigmap Bull. 28 (1980) 39–41.
20. C.A. Haverly, Behaviour of LP models, ACM Sigmap Bull. 22 (1977) 52–60.
21. C.A. Haverly, Studies of the behaviour of recursion for the pooling problem, ACM Sigmap Bull. 25 (1978) 19–29.
22. C.A. Haverly, Behaviour of recursion model More studies, ACM Sigmap Bull. 26 (1979) 22–28.

23. IChemE, A guide to capital cost estimating, 3rd ed., IChemE and The Association of Cost Engineers, London, 1988.

24. C.J. King, Separation Processes, 2nd ed., McGraw-Hill, Inc., New York, NY, 1980.

25. A.P.R. Koppol, M. Bagajewicz, B.J. Dericks, M.J. Savelski, On zero water discharge solutions in the process industry, Adv. Environ. Res. 8 (2003) 151–171.

26. Kuo, W. J. 1996. "A combined approach to water minimisation and effluent treatment system design", Ph.D. dissertation. Department of Process Integration. UMIST, Manchester, UK.

27. W.J. Kuo, R. Smith, Effluent treatment system design, Chem. Eng. Sci. 52 (23) (1997) 4273–4290.

28. W.J. Kuo, R. Smith, Designing for the interactions between water-use and effluent treatment, Trans. IChemE 76 (Part A) (1998) 287–301.

29. [29] W.J. Kuo, R. Smith, Design of water-using systems involving regeneration, Trans. IChemE 76 (Part B) (1998) 94–114.

30. Metcalf, Eddy, Wastewater Engineering; Treatment Disposal and Reuse, 3rd ed., Mc Graw Hill, New York, 1991.

31. F. Palacios-Gomez, L. Lasdon, M. Engquist, Nonlinear optimization by successive linear programming", Man. Sci. 28 (10) (1982) 1106–1120.

32. K.P. Papalexandri, E.N. Pistikopoulos, A. Floudas, Mass exchange networks for waste minimization: a simultaneous approach, Trans. IChemE 72 (Part A) (1994) 279–294.

33. Perez Huerta, S. 1995. "Capital and operating cost trade-offs in wastewater minimisation".MScDissertation, Process Integration Department,UMIST, Manchester.

34. J.W. Porter, J.H. Blake, R.T. Milligan, Zero discharge of wastewater from petroleum refineries, in: Complete Water Reuse: Industry's opportunity. Based on papers presented at the National Conference on CompleteWater Reuse, April 23–27, 1973, pp. 448–457.

35. D. Prakoptol, T. Srinophakun, GApinch: genetic algorithm toolbox for water pinch technology, Chem. Eng. Process. 43 (2004) 203–217.

36. I. Quesada, I.E. Grossman, Global optimization of bilinear process networks with multicomponent flows, Comp. Chem. Eng. 19 (12) (1995) 1219–1242.

37. M. Sittig, Petroleum Refining Industry Energy Savings and Environmental Control, 1st ed., Noyes Data Corporation, New York, 1978.

38. R. Smith, Chemical Process Design and Integration, 2nd ed., John Wiley & Sons, 2005.

39. J. Viswanathan, I.E. Grossman, A combined penalty function and outerapproximation method for MINLP optimization, Comp. Chem. Eng. 14 (7) (1990) 769–782.

Chapter 2

The Strategic Importance of Teaching Operations Research for Achieving High Performance in the Petroleum Refining Business

Marcel Joly[a, b, c], Roger Rocha[d], Luiz Carlos F. Sousa[e], Marcia T. Takahashi[e], Pierre N. Mendonça[e], Leonardo A.M. Moraes[e], and André D. Quelhas[f]

[a]Centre of Excellence for Applied Industrial Automation Technology (CETAI), PETROBRAS S.A., São Paulo, SP 05508-970, Brazil

[b]Capuava Refinery (RECAP), PETROBRAS S.A., Mauá, SP 09380-904, Brazil

[c]Department of Chemical Engineering, University of São Paulo, São Paulo, SP 05508-970, Brazil

[d]Exploration and Production, Espírito Santo Operational Unit (E&P/UO-ES), PETROBRAS S.A., Espírito Santo, ES 29057-550, Brazil

[e]Operations Research Division, Corporate Information Technology (TIC), PETROBRAS S.A., Rio de Janeiro, RJ 20020-004, Brazil

[f]School of Refining Technologies, PETROBRAS Corporate University, PETROBRAS S.A., Rio de Janeiro, RJ 20211-230, Brazil

ABSTRACT

Introducing a good educational program in the industrial environment may importantly differ from implementing a well-taught course in academic settings in a number of aspects. Potential troublesome aspects should be identified in advance such that the learning environment can properly tackle those issues through specialized or innovative ways. In this paper, we highlight the key role of Operations Research (OR) for the modern, high-performing oil industry and describe the in-house implementation of a pioneer, corporate-wide strategic learning program on OR techniques at PETROBRAS, a semi-public Brazilian multinational energy company.

The learners were chemical engineers working at the planning/scheduling department of oil refineries, which has a high profile in the organization. Under tutoring, they acquired theoretical knowledge and practical skills concerning the development of optimization models, solution techniques and finished OR-based tools in a community of OR experts, thereby enabling them to rehearse their reflective and computational skills related to day-to-day activities. Most importantly, since aimed to foster the development of self-regulated, reflective learners, this course technically enabled and motivated planners and schedulers to develop further specialization, to rethink inefficient work processes, to change the attitudes and culture in the refinery, to question previous assumptions and operational paradigms.

INTRODUCTION

The increased efficiency of modern computer technologies allied to the development of powerful optimization algorithms (Viswanathan and Grossmann, 1990, Drud, 1996 and IBM, 2012) for solving large-scale decision-making problems have amplified the possibilities of mathematical modeling approaches (Kelly and Forbes, 1998, Kondili et al., 1993 and Pantelides, 1994) in an unprecedented way. In this context, Operations Research (OR) techniques (Hillier and Lieberman, 2005) and large-scale optimization (Biegler and Grossmann, 2004) have proven to be instrumental to achieve important progress in many fields of science and technology, with innovative solutions being successfully implemented on a myriad of real-world problems, from corporate-wide supply chain management (Neiro and Pinto, 2004) to highly specialized biomedical and molecular biology systems (Joly and Pinto, 2006 and Joly and Odloak, 2013).

Particularly in Brazil, the energy sector has experienced an astonishing change in the last 40 years and PETROBRAS, a semi-public Brazilian multinational energy company, must be prepared to cope with new multidisciplinary challenges to sustain its competitiveness, since it deals with one of the world's longest and most complex supply chains (Iachan, 2009), beginning with a natural resource in the ground and continuing all the way through to the end-users. As any other global energy company, PETROBRAS has been forced to search for better operating points locally as well as to overstep their physical boundaries in order to visualize the surrounding business environment before planning their internal activities (Neiro and Pinto, 2004). In addition, since the Brazilian oil sector monopoly breakdown in 1997, the PETROBRAS downstream business model is continuously moving toward a stiff competition demand-driven model rather than the conventional monopoly supply driven one. At the same time, a number of new, multidisciplinary, constraints arise: increase in oil prices leading to the need for processing poor quality crudes, increasing demand and new demand patterns for petroleum products including petrochemicals, new Brazilian stringent environmental regulations for clean fuels, start-up of new production technologies (e.g., heavy ends upgrading, product quality improvement, efficient fuel usage, controlling refinery emissions, in-line blending units) embedded into more complex refining schemes and subject to the increasing complexity of the logistic hardware (Rejowski and Pinto, 2003) that interconnects Brazilian crude oil terminals, refineries, and distribution centers, a problem exacerbated by the need of implementing lower supply chain inventory levels (Magatão et al., 2011). This nervous scenario is further subject to possible disruptions caused by various factors at the global level that are hard for any oil company to control.

In fact, the oil industry has represented one of the most prolific fields for the application of OR techniques since the 1950s (reviewed in Cooper, 2002). Typically part of a complex supply chain, oil refineries convert raw-materials into several finished products aiming at achieving customer satisfaction, and the company's image depends largely on a reliable and profitable production planning and scheduling committed to and in compliance with customer needs timely (Joly, 2012). Besides performance optimization, intelligent refinery production planning and scheduling allow forecasting and, hence, *safety* from an operational point of view. For instance, avoiding environmental disasters became vital for the survival of any oil company in an ever-increasing competitive global market (Zhang and Valleur, 2010).

In order to support the oil refineries in achieving these goals, a number of excellent commercial and proprietary solutions dedicated to support strategic (long-term), tactical (medium-term) and operational (short-term) decision-making processes have been developed (Table 1). Whereas commercial technology has provided converged solutions for refinery production planning and scheduling problems since the 1970s (Bonner and Moore, 1979), currently there is a possibility of going beyond what was commercially available aiming at exploiting new frontiers (Agrawal and Balasubramanian, 2006, Guerra et al., 2010 and Swensen and Acuff, 2011): specialized, more accurate and integrated optimization tools (Maravelias and Sung, 2009).

Table 1: Examples of optimization software for refinery operations

Refinery	Planning	Refinery	Scheduling
Company	Product	Company	Product
Aspen Technology	PIMS	Aspen Technology	Aspen Petroleum Scheduler
Invensys	Spiral Plan	Invensys	Spiral Schedule
Haverly Systems	GRTMPS	Haverly Systems	H/SCHED
Industrial Algo-rithms	IMPL[a]	Industrial Algorithms	IMPL
Prometeus	SIMRAF	Prometeus	PROLAV
Honeywell	RPMS	Princeps	Flowers
ExxonMobil	Platoform	Soteica	Visual Mesa
Shell Global Solu-tions	GEMMS	Technip	Forward C
Ecopetrol	SIGMA	Petrobras	BR-SIPP

[a]IMPL is a modeling platform for process industries.

In this context, a number of real-world industrial OR-based applications for challenging problems have been developed and implemented as a result of tight collaborations between PETROBRAS and the academy (e.g., Moro et al., 1998, Pinto et al., 2000, Joly and Pinto, 2003, Rejowski and Pinto, 2003,Magalhães, 2004, Neiro and Pinto, 2004, Micheletto et al., 2008, Rocha et al., 2009, Iachan, 2009,Magatão et al., 2011, Aires et al., 2011, Magatão et al., 2011, Joly, 2012, Feital et al., 2013, Mendoza et al., 2013, Menezes et al., 2014a, Menezes et al., 2014b, De Souza et al., 2010 and Moro and Grossmann, 2013). In addition, the introduction of commercial (e.g., PIMS in the 1990s) and in-house developed OR tools (e.g., Magalhães et al., 1998) have revealed the opportunity for PETROBRAS refineries to rethink current solution approaches, working processes and operational – or even strategic – paradigms (discussed in Joly, 2012). For instance, *"satisfying the foreseen demand for oil products in Brazil"* is a sexagenarian hard constraint from the monopoly era that remains valid for the PETROBRAS planners when they run corporate or local planning models. While a semi-public energy company and the principal oil refiner in Brazil, the PETROBRAS public image remains largely associated with its ability to meet the Brazilian fuel market needs, which now include biofuels.

On the other hand, the increasing focus on a safe, sustainable and profitable performance has promoted a culture change on how the modern refining business should be managed in Brazil. A clear understanding emerged concerning the fact that front-edge technology alone is not a sufficient condition to exploit the potential benefit promised by the modern optimization technology. Rather than sophisticated computational tools, necessary conditions to stretch refinery operation involve harmonic working processes performed by a highly educated and qualified staff (Zhang and Valleur, 2010).

Such staff then becomes able to efficiently operate advanced refinery automation tools, which must be still properly integrated with huge information systems and corporate databases formally identified in the business-planning process of the company in order to capture the existing economic opportunities (Iachan, 2009). Strictly speaking, the development of such kind of expertise is a long process, during which theoretical, practical and metacognitive elements of expert knowledge are integrated into a coherent whole (Katajavuori et al., 2006).

The educational agenda discussed in this paper addresses a real key question concerned with the fact that employers in the chemical process industry are currently demanding a greater level of competencies from engineers: versatility in a range of areas, not just the core technical domain (Gomes et al., 2006). Among these competencies is OR which should be understood as a critical technology for achieving sustainable competitiveness in a globalized economy. In particular, this paper means to provide some guidelines about how OR can efficiently be taught in the industrial environment to experienced chemical (or more generally,*process*) engineers daily involved with refinery production planning and scheduling activities and how the acquisition of this expertise by a high-profile staff within the corporation can add value to the performance of high-performing oil refineries. This training program was designed by the PETROBRAS Centre of Excellence for Applied Industrial Automation Technology (CETAI) and the PETROBRAS Corporate University (PCU) in tight collaboration with other OR expert company's areas. The challenge here was to condensate a large and complex theoretical background on large-scale optimization and mixed-integer programming (MIP) techniques into a concise and didactical content to stimulate the learners and to give them: (a) core competencies in mathematical optimization to properly handle more sophisticated computer-aided decision-making tools, (b) the basis on MIP for modeling and solving ever-increasing complexity decision-making problems in a commercial modeling system, and (c) the foundations for reflexive learning as well as critical analysis on work processes in a dynamic workplace.

This paper is organized as follows. Section 2 describes the course participants, methodology and structure. The educational content is described in Section 3. Results are discussed in Section 4 whereas Section 5 concludes the paper.

COURSE DESCRIPTION

Participants

The participants of this learning program were mostly (but not restrict to) process engineers not familiarized with large-scale optimization. These learners, who were mostly young engineers with no more than 5 years of industrial experience in refinery production planning and scheduling, came from several areas of PETROBRAS, including petrochemical and oil refineries (mainly), research and development (R&D) centers and headquarter offices. The remaining learners were experienced IT and OR analysts without expertise in MIP, involved with the in-house development and maintenance of OR tools. These learners have multidisciplinary undergraduate education.

Methodology

An important challenge of higher education is to integrate the acquisition of domain-specific knowledge with practical knowledge and skills and, further, to encourage learners to apply knowledge, to solve problems and to tackle problems at higher levels (Tynjälä et al., 2003). The significance of practical training in linking theoretical studies with practice was investigated in Katajavuori et al. (2006), which concluded that it would be important to use practical learning more effectively to support theoretical studies, as the practical training may foster learner's learning and linking theory and practice. Particularly during the beginning of the first two modules (LP and MILP), short periods of practical studies based on computer-assisted learning were carried out with tutor support combined with theoretical studies. The tutoring structure also differed at each module as learners did not have an individual tutor assigned, rather 3–5 tutors who circulated around all the learners. Here, classic OR toy problems were solved. This method offered the learners more opportunities to apply their knowledge and to train their skills.

In module 4 (MINLP), a complex, real-scale problem concerning refinery short-term scheduling was addressed and distinct instances with increasing complexity were solved. Basic topics in computer architecture were also addressed in order to improve learner motivation toward developing a finished optimization solution which can be operated through a graphical user interface (GUI) by non-expert personnel. In this sense, given the wide use of spreadsheets for refinery production planning and control activities, a Microsoft Excel®-based GUI, which can also be easily integrated to enterprise databases, was then fully developed by the learners for the most complex problem instance studied.

When planning the structure of the course, we also had to consider the constrained availability of time of the learners, most of them heavily demanded refinery›s production planners and schedulers. Since this staff is permanently involved with specialized supply chain decision-making processes (some of whom in real-time) affecting refinery profitability and operability, any long-term training program conducted outside the refinery may be unfeasible from a practical point of view. Thus, a modular learning program carried out at the PETROBRAS Corporative University was carefully planned aiming at minimizing the impact of the staff absence in their workplace at specific periods of each month. Clearly, since one of the behavioral characteristics of learners indicates that not only the rate of what learners learn must be accounted for, but also the rate of decay that takes place after an objective is mastered must also be taken into account (U.S. Department of Defense, 1975), reinforcement loops were built into the instructional process associated with each new module. This was also convenient due to the complexity of the MIP subject and to the variety of abilities among the learners including differences in educational background and professional experience.

Tools

Whereas spreadsheet solvers and widespread engineering software (e.g., MATLAB®) may be advantageous to address small-size problems regarding didactical ends (Ferreira et al., 2004), process engineers typically have to cope with more complex systems along their careers and advanced mathematical modeling and solution technology are required (García et al., 2012). At the educational level, solving real-world process systems engineering optimization problems can be made easier to the learners by using available algebraic modeling systems (Table 2), whose features includes powerful high-level languages, this mode offering a straightforward way to implement, in a compact way, complex optimization models of varying types, which may be either linear (Dantzig, 1963) or nonlinear (Bazaraa et al., 1993) in nature and with continuous, discrete (Nemhauser and Wolsey, 1993) or both types of variables.

Table 2: Examples of algebraic mathematical programming languages

Company	**Product**
AMPL Optimization	AMPL
DASH Optimization	MP-Model and MOSEL
GAMS Development	GAMS
IBM	OPL
LINDO Systems	LINGO
Paragon Decision Technology	AIMMS
Princeton University	MINOPT

GAMS (Brooke et al., 1998), which has the advantageous feature of easy interfacing with conventional spreadsheets and engineering software, has successfully been applied in teaching engineering by many authors (Ferris, 1998, Mingo et al., 2011 and García et al., 2012). Here, we also adopted GAMS as the solution platform to implement the optimization problems proposed and their solution methods. As a reference, each course participant received a classical OR textbook (Hillier and Lieberman, 2005) donated by PCU at the beginning of the course.

Learning Scope and Structure

In spite of the current importance of OR as an educational subject for chemical engineers, there has been a limited number of publications related to the teaching or assessment of courses related to large-scale optimization (García et al., 2012). Here, foundations on linear (LP), nonlinear (NLP), mixed-integer linear (MILP) and mixed-integer nonlinear (MINLP) programming and meta-heuristics (MH) were taught separately in 6 learning modules along a 9-month period according to scope, objectives and workloads informed in Table 3. In any complex course, a major step in the design phase is to determine the program sequence and structure to ensure the learning objectives are met. Proper sequencing also helps to avoid inconsistencies in the content of instruction and provides the learners with a pattern of relationship so that each activity will have a clear purpose (U.S. Department of Defense, 1975). By extending LP concepts introduced in the first module, MILP was then addressed in the sequence. For such a strategy, two major advantages were clearly identified. Firstly, the discussion on simple, nevertheless illustrative, optimization problems involving discrete decisions (i.e., integer variables) early in the training program was highly motivating for the learners, since they could promptly visualize the potential applicability of mathematical programming techniques to real-world problems of their main interest: those problems concerning a lot of combination possibilities for resource selection and sequencing throughout a time period of interest, that is, short-term scheduling problems. NLP was introduced prior to MINLP (Module 4), which addressed the most complex (and non-deterministic polynomial-time hard (NP-hard) (Vavasis, 1991)) exercises. An additional module dedicated to the development of a conventional GUI in Excel for the most complex optimization problem studied was also proposed to enable learners to locally and quickly develop “finished” specialized MIP-based solutions for particular optimization problems they are faced with daily in the working environment, which may fail to be addressed or not properly represented by other general-purpose OR solutions already implemented at the corporative level. Opportunities here may include a more detailed/accurate modeling of critical systems in the sense of process representation, decision variables and/or performance criteria, or feasibility studies concerning alternative/future configurations of the system and its surrounding conditions. Since OR solutions, and particularly those related to short-term scheduling, usually involve managing a large set of data, the development of an integrated GUI for such customized optimization solutions may be interesting in the sense of allowing faster use and to enable non-MIP-expert engineers to operate them.

Table 3: The training program structure

	Topic	Intended objectives	Scope	Work-load (h)
1	LP	To learn the foundations of LP. To acquire familiarity with the use of a general algebraic modeling system and LP-based algorithms. To be able to model, to implement and to solve OR toy-problems and to interpret the solution output from a professional modeling platform.	Theory, 11 exercises and introduction to GAMS	40
2	MILP	To learn the foundations of combinatorial optimization. To learn modeling good-practices in MIP and to apply them by solving small discrete optimization problems related to the refinery operation decision-making.	Theory and 11 exercises	32
3	NLP	To learn the foundations of NLP and to apply it by solving simple optimization problems of systems process engineering.	Theory and 3 exercises	24
4	MINLP & Real-world applications	To be familiar with a well-established solution method for MINLP problems. To learn linearization techniques. To experience the major difficulties associated with modeling and computational implementation of large-scale MINLP problems. To be aware of a hierarchical decision-making framework for the management of the refining business (see Fig. 5).	Theory and 2 exercises	32
5	GUI	To learn how to interface GAMS with popular electronic spreadsheet (Microsoft Excel) thus enabling the learners to develop a user-friendly "finished" optimization tool, which can be operated by non-experts on MIP or GAMS.	1 exercise	16

6	MH	To introduce the learners to alternative solution approaches aimed at quickly producing 'good' feasible solutions for large-scale combinatorial optimization problems.	Theory and 2 exercises	16
			Total	160

Overview of Module 1 (LP)

The largest course workload was devoted to providing the learners with an in-depth view of LP. Initially, OR definition and some classic OR applications were presented. Historical aspects of OR efforts since the 1960s at PETROBRAS (Iachan, 2009) were outlined and current challenges in OR developments were reviewed in the light of the company's strategic plan for the period 2015–2030. This introductory section of the course was concluded by discussing general work phases embedded in developing a solution for any complex decision-making problem: *problem understanding* (as the crucial step for achieving success),*variables mapping, system modeling, solution testing and validation,* and the usual *development of an information system for automatic solution operation* (data input/output). The LP theory was then focused on. Specific LP applications in the oil industry since the 1950s (Symonds, 1955) were concisely reviewed before introducing the theoretical background (Table 4). Basic information and concepts were initially introduced by lecture-based learning. Classical and refinery LP toy problems were considered along with the theoretical content of the lectures (Table 5).

Table 4: Module 1 content

Topic	Scope of learning	Solved examples[a]	Proposed exercises[a]
LP concepts and modeling	LP problem structure and basic definitions (e.g., linearity, continuity, convexity, feasibility, optimality). Introduction to problem modeling. Guided and proposed exercises. Overview of strengths and weaknesses of commercial and open-access modeling tools.	i	–
Graphical solution	Graphical representation of the feasible region in R^2, gradient vector of the objective function value (OFV) and the OFV curves. Graphical identification of single and multiple global optimal solutions, infeasible problems, and unbounded problems.	i	–
Algorithms	The Simplex method. Graphical interpretation, strengths and weaknesses of the method. Overview of the Interior Point method.	ii	–
Duality	Motivation. Formulation. Properties of a dual formulation. Primal and Dual Simplex methods. Basic concepts (e.g., marginal value).	i, iii	–
Sensitivity analysis	Motivation. Basic definitions. Post-optimality analysis. Graphical interpretation.	i	–
GAMS	Modeling platforms. Overview of GAMS modeling system and its features. Solved example based on the GAMS tutorial.	iv	v to x
Assessment	Modeling, implementing and solving an LP problem	–	xi

[a]Solved examples and proposed exercises themes are summarized in Table 5.

Table 5: Summary of exercises

#	a	Subject
i	*	Determining the optimal crude blend plan for a crude distillation unit subject to technical (product yields) and commercial (foreseen demand) constraints, and an economic objective.
ii	*	Min (30x + 40y), subject to: 3x + 2y ≤ 60, x + 2y ≤ 45, x + y ≥ 10, x ≥ 0, y ≥ 0.
iii	*	Determining the optimal logistic plan for fuel distribution from n terminals to m customers subject to technical and commercial constraints, and an economic objective.
iv	*	The transportation problem adapted from Dantzig (1963).
v	*	Processing capacity sizing of a new oil refinery subject to technical and commercial constraints, and an economic objective.
vi	*	Determining the optimal blend recipe for a computer-controlled gasoline in-line blending unit subject to technical and commercial constraints, and an economic objective.
vii	*	The same as i, but including additional technical constraints (bounds on crude blend composition).
viii	*	Determining the optimal production planning for a simplified refining scheme subject to technical (processing capacities, product yields and qualities) and commercial (foreseen demand) constraints, and an economic objective.
ix	*	The same as vi, but including additional technical (availability, quality and cost of each blending component, andspecification bounds for finished products) constraints.
x	*	Determining the optimal supply chain plan for an oil company subject to technical (refining capacity) and commercial (foreseen demand) constraints, and an economic objective.
xi	*	The same as viii, but considering a refining scheme of higher complexity and an expanded set of finished products and physical–chemical properties (or qualities) to be managed.

[a]Relative complexity level: *green*: low complexity; *yellow*: intermediary; *red*: high.

Initially, a guided example that can be represented in R^2 related to crude blend selection is solved aiming at introducing important concepts of LP. Literal modeling is firstly derived for representing the problem constraints and then translated into mathematical formulations. The resulting mathematical constraints are graphically represented in R^2 such that the problem feasible region can be visually identified. Convexity properties are then discussed and the global optimal solution is graphically determined. The generic structure of a LP problem is then presented and concisely posed in its matrix form. Particular cases, such as problems with multiple optimal solutions, infeasible or unbounded problems (typically an outcome from modeling errors), are discussed on the basis of modified versions of the guided example *i* and graphically interpreted. Two LP algorithms were discussed. Firstly, a fine-grained view of the Simplex method (Dantzig, 1963) was presented and its solution rationale graphically illustrated step-by-step based on a guided example in R^2. After discussing major strengthens and weaknesses of this method, the Interior Point method, a polynomial time algorithm for solving LP proposed by Karmakar (1984), was outlined.

Duality was introduced and mathematically formalized. The motivation for post-optimality analysis in LP was discussed by considering the deterministic nature of the method and uncertainties of the real-world. A practical view of how performing sensitivity analysis in LP occurred based on guided example *i*, for which a number of realistic "what-if" questions were posed, modeled, and graphically interpreted in order to clarify key concepts (e.g., active/inactive constraints, marginal values, break-points, etc.) and analytical possibilities. Guided examples were solved in MS-Excel by the learners. Typical interpretation mistakes of dual variables meaning by non-experts in LP were pointed out (e.g., quantifying opportunities based on marginal values far from the optimal solution). Illustrative real-world examples of sensitivity analyses based on the PETROBRAS LP planning model for corporate investments in the long-term were presented. Based on the GAMS User's Guide (Brooke et al., 1998) content, a customized tutorial on GAMS was elaborated aiming at providing a quick but complete overview of GAMS and its major features. Proposed exercises were implemented and solved in class by the learners. Concluding this introductory module, a final exercise was solved by the learners in class for assessment ends.

Overview of Module 2 (MILP)

Firstly, the motivation for discrete optimization was carried out in a one-hour class. Critical aspects on modeling and solving any complex decision-making problem were initially addressed (Fig. 1). Mathematical programming classes (LP, MILP, NLP and MINLP) applications were characterized in the light of typical refining problems and modeling approaches usually considered for representing them. Here, illustrative problems related to refinery planning (Tucker, 2001), crude oil scheduling (Lee et al., 1996 and Kelly and Mann, 2003), gasoline and diesel blending (Barsamian, 2007 and Castillo et al., 2013), and production scheduling of liquefied petroleum gas, fuel oil and asphalt (Pinto et al., 2000 and Joly and Pinto, 2003) were addressed. Technical distinctions between production planning and scheduling were highlighted in terms of complexity issues and business aims (Fig. 2).

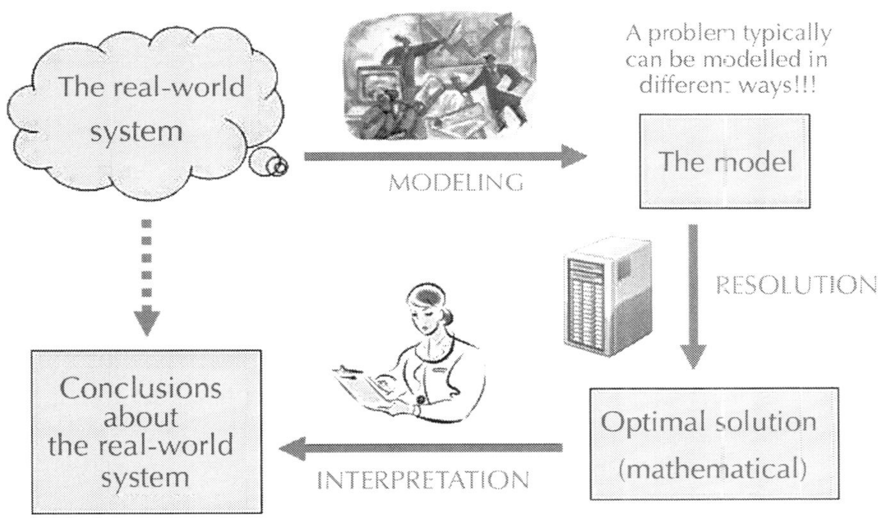

Figure 1: The process of solving a real-world decision-making problem.

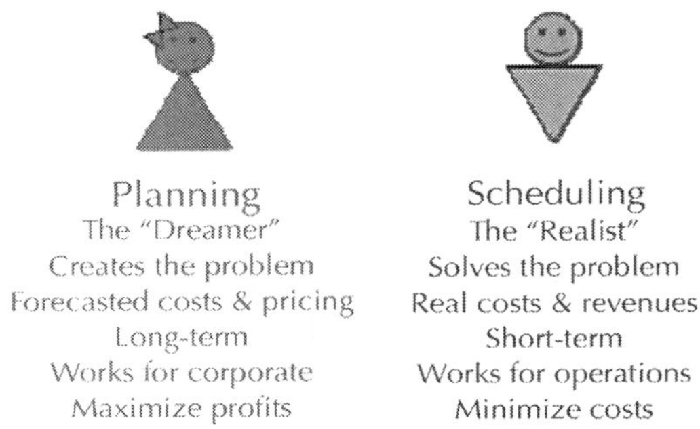

Figure 2: Planning vs. scheduling.

A short review of LP was presented in order to refresh the learners' memory (there was a gap of about one month between modules 1 and 2) and to integrate or to relate the previously taught content with the new topic (Table 6). Basic concepts of a number of algorithms for solving discrete problems (e.g., Bellman, 1957, Gomory, 1960, Land and Doig, 1960 and Benders, 1962) were discussed in terms of their major characteristics. A more detailed view of the branch and bound method was then presented by considering a simple MILP problem and constructing the associated binary tree. Guided exercises *xii* and *xiii* were solved. Here, simple integer decisions embedded in specialized business rules (the model constraints) were modeled step-by-step in order to familiarize the learners with discrete-continuous modeling. The translation of logical conditions into mathematical constraints (Raman and Grossmann, 1991) was exemplified with respect to typical operational rules in the refining business (e.g., blending operations integrated with refinery dispatch logistics). GAMS special language features (variable types, setting variable priorities, modeling scaling) were presented and major GAMS and CPLEX options for MIP problems were discussed in depth. Classic OR problems were then solved (examples *xiv* to *xviii*). Two major MILP algorithms were discussed in detail. Improved modeling in MILP (Kelly and Zyngier, 2007) and general good modeling practices in MIP were demonstrated in practice by solving example *xix*. Finally, four exercises of higher complexity (exercises *xx* to *xxiii*) were solved in class by the learners. In order to optimize the use of the available time by eliminating "mechanical", time-consuming, tasks by the learners, pre-formatted GAMS files (including model scalars, sets and parameters for each exercise) were prepared by the instructor and handed out in advance to the learners, thereby allowing them to focus on the problem modeling. By doing this, modeling nomenclature standardization among the learners was ensured and typing errors with regard to the problem input data avoided (Table 7).

Table 6: Module 2 content

Topic	Scope of learning	Solved examples[a]	Proposed exercises[a]
Introduction	Motivation for using decision-making technology in the oil industry. Overview of mathematical programming techniques. Revision in LP. Why using MILP?[b]	xii, xiii	–
MILP overview	Integer and binary variables. The MILP problem formulation. Solved examples in classical OR problems. Ensuring model consistency. Good modeling practices in MILP. Logical inferences.	xiii [c] xiv to xviii	–
Algorithms	Complexity issues. Major MILP algorithms. Basic concepts in MIP. The branch and bound method. The binary tree and search strategies. Cut plane methods. Good practices in MIP. Commercial solvers for MILP. Solver parameterization (overview of CPLEX tuning parameters).	xix	–
Applications	Interpreting, modeling, implementing and solving MILP problems.	–	xx to xxii
Assessment	Interpreting, modeling, implementing and an MILP problem.	–	xxiii

[a]Solved examples and proposed exercises are summarized in Table 7.

[b]Here, illustrative examples were solved in R^2 to show that [0,1] variables in the optimal solution of an LP model cannot be "manually" approximated to {0,1} values aiming at producing discrete decisions and that a new rigorous approach is, therefore, required.

[c]A modified version with higher complexity was formulated to illustrate the use of logical inferences.

Table 7: Summary of exercises

#	a	Subject
xii	*	Determining the best quantity of each automobile type to be produced by an automotive industry subject to commercial constraints (bounds on foreseen demands) and an economic objective.
xiii	*	(The knapsack problem) Selecting the best set of oil wells to be drilled subject to budgetary constraints and an economic objective in terms of the foreseen profitability associated to each oil well.
xiv	*	(The knapsack problem) Selecting the best set of revamp projects for an oil refinery to be implemented subject to commercial constraints (budget) and an economic objective in terms of foreseen profitability associated to each revamp proposal.
xv	*	Selecting the best painting sequence to be considered by an automotive industry to paint distinct lots of cars with different colors subject to operational constraints (setup times for painting equipment cleaning) and a technical objective (minimizing the makespan).
xvi	*	(The generalized assignment problem) Determining the best assignment among tasks and machines subject to operational constraints and an economic objective.
xvii	*	(The capacity location problem) Determining the best placement of fuel distribution centers regarding the required investments (as a function of localization) and transportation costs subject to operational and commercial constraints and an economic objective.
xviii	*	(The job shop scheduling problem) Determining the best processing sequence for jobs subject to operational constraints and a technical objective (minimizing the makespan)

xix	•	(The Big M formulation) Determining the optimal number of items that should be produced by a factory considering that there is a minimal lot size, fixed and variable operational costs, and an economic objective.
xx	•	(The cutting stock problem) Determining the best cut plan for cutting paper rolls subject to commercial constraints (demands by fixed roll widths) and an economic objective (minimizing either the number of paper rolls to be processed or the waste).
xxi	•	Determining the best operation schedule of crude tanks and operation modes to crude distillation units regarding set-up costs, and an economic objective.
xxii	•	Determining the best logistic plan for multiregional markets supply by a set of refineries subject to operational and commercial constraints and an economic objective.
xxiii	•	Determining the best refining scheme for a refinery faced by direct competitors from whom intermediary streams can be bought instead of produced subject to operational and commercial constraints and an economic objective.

[a]Relative complexity level: *green*: low complexity; *yellow*: intermediary; *red*: high.

Overview of Module 3 (NLP)

Before introducing nonlinear programming, we revisited basic concepts of linear algebra in a two-hour class. Fundamental concepts in nonlinear optimization (e.g., convexity and nonconvexity, global and local optimality, optimality conditions, etc.) were introduced and graphically illustrated in R^2. The general structure of a nonlinear optimization model was stated and NLP problems were categorized into major classes: unconstrained optimization, quadratic programming, (non)convex programming, geometric programming, separable programming. In order to illustrate typical algorithm difficulties in solving NLP problems, LP and NLP-based models were graphically solved in R^2. Here, two types of NLP models were considered. The first type considered a linear objective function model and nonlinear constraints whereas the second one was based on a nonlinear objective function model and linear constraints. Good modeling practices were discussed. Here, practical examples were formulated and presented to show

how one can: (a) provide good initial search points (in order to reduce the variance in derivatives (Jacobian elements) associated with the initial point), (b) implement linearization techniques (by approximation) in order to reduce the model nonlinearity, (c) implement smoothing techniques for non-differentiable functions (e.g., maximum and minimum functions, Euclidian norm) with a practical application (representation of a step in flowrate profile), determine bounds on variables, perform model scaling, etc. NLP algorithms were then addressed. Firstly, one-dimensional search and the Newton methods were formalized and simple examples of unconstrained nonlinear optimization were presented and solved graphically. A detailed discussion on first-order necessary conditions (Karush–Kuhn–Tucker) and second-order conditions for local optimality were presented and illustrated in R^2 before introducing algorithms for constrained optimization. Major commercial technology for solving NLP problems regarding the generalized reduced gradient (GRG), sequential quadratic programming (SQP) and interior point method or barrier method were presented. Strengths and weaknesses of commercial large-scale NLP solvers were then systematically discussed in terms of the model structure (e.g., convexity, presence of quadratic functions, problem size, number of freedom degrees and degree of "nonlinearity" of the model) (Table 8 and Table 9; Fig. 3).

Table 8: Module 3 content

Topic	**Scope of learning**	**Solved examples**[a]	**Proposed exercises**[a]
Introduction	Overview of NLP applications in the oil industry. Review of linear algebra.	–	–

Unconstrained optimization	Optimal solution characterization. Graphical interpretation in R^2. Non-differentiable problems. Algorithms: graphical interpretation, strengths and weaknesses of each method.	xxiv	–
Constrained optimization	Optimal solution characterization. Graphical interpretation in R^2. Non-differentiable problems. Quadratic problems. Algorithms: graphical interpretation, strengths and weaknesses of each method. Commercial solvers for large-scale optimization (strengths and weaknesses in terms of the optimization model structure).	xxiv	xxv, xxvi
Modeling good practices	Linearization techniques (by approximation). Initial search point. Setting variable bounding and dominion. Model scaling. Smoothing techniques.	xxiv	–
Assessment	Modeling, implementing and solving an NLP problem	–	xxvii

[a]Solved examples and proposed exercises themes are summarized in Table 9.

Table 9: Summary of exercises

#	a	**Subject**
xxiv	*	Unconstrained and constrained NLP-based models were considered here aiming at focusing on the key issues of the learning scope summarized in Table 8.
xxv	*	Determining the optimal design of a pressure vessel subject to the American Society of Mechanical Engineering (ASME) standards and material costs and an economic objective.

xxvi		Determining the best parameter fit for a literal flash point correlation[b] given a large set of experimental data. The objective is to minimize the overall quadratic error between experimental and calculated data.
xxvii		(The pooling problem – see Fig. 3) Determining the best price policy for products[c] of an oil refinery subject to technical (raw-material qualities and refinery processing hardware) and commercial (foreseen demand and product specification) constraints. The objective is to maximize the refinery receipt with respect to two fuel oil grades.

[a]Relative complexity level: *green*: low complexity; *yellow*: intermediary; *red*: high.

[b]The Chevron flash point method was considered: $I = \exp((k_1/(pf + k_2)) - k3)$. Here, I = flash point index, pf = flash point temperature (°F).

[c]Here, it is assumed that the product price is a function of its contaminant (sulfur) content: $p = p_0(2 - (s_1/s_2))$, where p = product price per volume unit,p_0 = baseline product price, $s1$ = actual sulfur content in the product, and s_2 = upper bound for sulfur content (product specification).

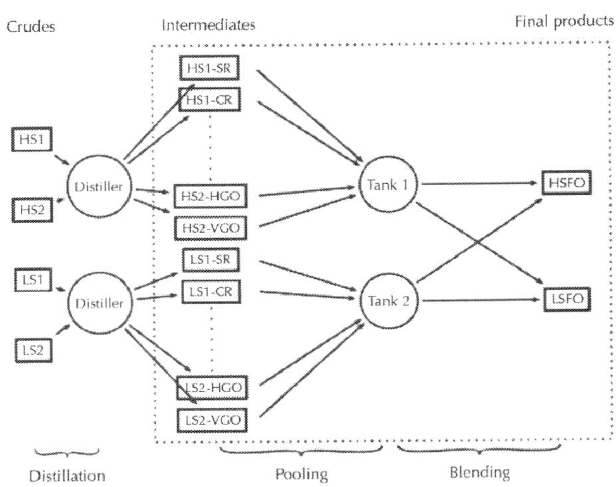

Figure 3: Schematic representation of the pooling problem (exercise *xxvi*, Table 9). *Abbreviations*: HS: high sulfur crude; LS: low sulfur crude; SR:

straight run; CR: continuous run; HGO: heavy gasoil; VGO: vacuum gasoil; HSFO: high sulfur fuel oil; LSFO: low sulfur fuel oil.

Overview of Module 4 (MINLP)

In this module, we addressed OR applications with practical interest in oil refineries. The learning scope of this module comprised: (a) to obtain basic knowledge about planning and scheduling models; (b) to learn good modeling practices for solving large-scale combinatorial optimization problems, and (c) to deepen the knowledge on GAMS usage through implementing complex models.

Main topics and associated contents are summarized in Table 10. As a motivational topic, foundations of the oil industry supply chain management were initially reviewed. The hierarchical decision-making framework that governs the PETROBRAS downstream model (Moro, 2009 and Joly, 2012) was then detailed by focusing on the refinery production planning and scheduling business layers (Fig. 4). The role of planning and scheduling as integrator activities (Steinschorn and Hofferl, 1997) for work processes was discussed. Major strengths and weaknesses of some of the most worldwide implemented commercial solutions for supporting decision-making in the refining business (discussed in Joly (2012)) and in-house developed solutions (such as BR-SIPP (Magalhães et al., 1998) and PLANREF (Aires et al., 2011)) were pointed out in the light of current and future needs of planners and schedulers to have accurate, integrated, and optimization-based computer-aided decision-making easy-to-use tools with strategic, operational and economic metrics (Moro, 2003, Agrawal and Balasubramanian, 2006, Swensen and Acuff, 2011 and Joly, 2012).

Table 10: Module 4 content

Topic	Scope of learning	Solved examples[a]	Proposed exercises[a]

Motivation	Supply chain downstream overview. The hierarchical planning framework. Commercial software technology for refinery planning and scheduling.	–	–
Scheduling problems	Basic concepts. Problem classification. MINLP model formulation. The outer-approximation with equality relaxation algorithm. Good modeling practices.	xxviii toxxx	xxxi
Assessment	Modeling, implementing and solving a real-world refinery short-term scheduling problem. Model linearization.	–	xxxii, xxxiii

[a]Solved examples and proposed exercises themes are summarized in Table 11.

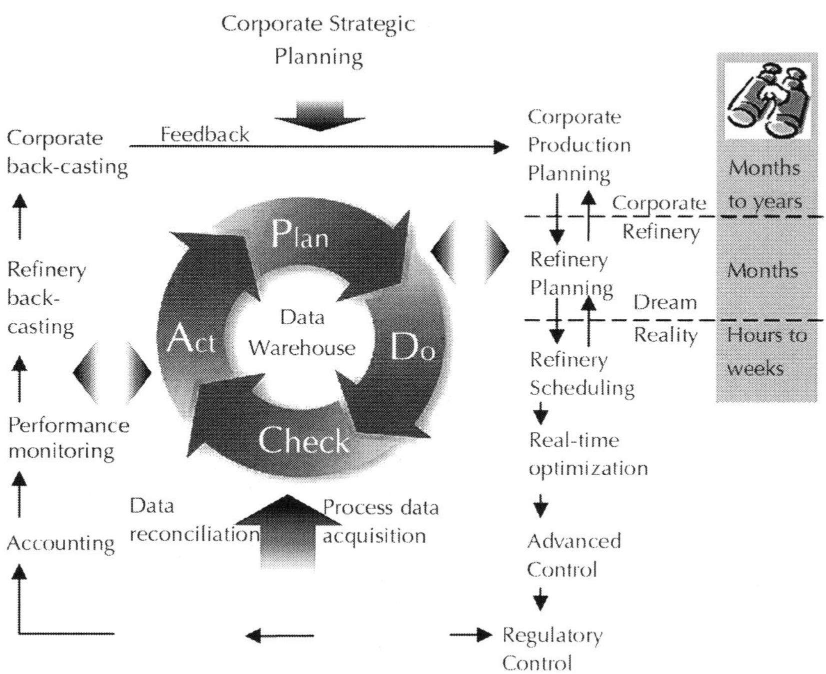

Figure 4: Schematic representation of a hierarchical decision-making framework in refining business.

Planning and scheduling problems were revisited with respect to basic definitions (Reklaitis, 1991), classification (Pinto and Grossmann, 1998), and complexity (Garey and Johnson, 1979). Problem representations using a traditional State Task Network (Kondili et al., 1993) or a Resource Task Network (Pantelides, 1994) were illustrated and discussed with respect to benefits and limitations. Practical exercises on good modeling practices (i.e., avoiding unnecessary non-convexities, models with binary variables, logical relationships, big-M reformulations, etc.) were then illustrated in examples *xxviii* to *xxx* (Table 11). The advantages and drawbacks associated with either a discrete or a continuous representation of time were discussed (see Joly and Pinto, 2003), by emphasizing that the underlying time framework to be considered is one of the major differences in the basic structure of MIP formulations used for production scheduling problems (Stefansson et al., 2011).

Table 11: Summary of exercises

#	[a]	**Subject**
xxviii	*	Determining the optimal production schedule for a general chemical plant from Kondili et al. (1993).
xxix	*	Determining the optimal logistic schedule for fuel distribution operations by fuel trucks from n terminals to mcustomers subject to technical and commercial constraints, and an economic objective.
xxx	*	Determining the optimal operation schedule of crude tanks and operation modes to crude distillation units regarding set-up costs, and an economic objective.
xxxi	*	Refinery short-term scheduling of crude oil processing with inventory management subject to technical and commercial constraints, and a technical objective.
xxxii	*	Refinery short-term scheduling of crude oil unloading with inventory management subject to technical and commercial constraints, and a technical objective. The "motivating example" from Lee et al. (1996).
xxxiii	*	"Example 4" from Lee et al. (1996) (see Supplementary Material).

[a]Level of relative complexity. *Green*: low complexity; *yellow*: intermediary; *red*: high.

The problem of refinery short-term scheduling of crude oil processing with inventory management (exercise *xxxi*; Fig. 5) was presented to the learners. The model instance considered 12 crude oil tanks, 2 crude distillation units (CDUs), 22 crude types, 8 physical–chemical properties to be managed over 3 CDU cuts of main interest (naphtha, diesel and atmospheric residue). Non-linear blending models were considered for the product quality specification calculations. A semi-finished GAMS execution file comprising the full model parameterization (initial volume and composition for crude tanks, crude assays, etc.) variable declarations was given to the learners for providing model nomenclature standardization previously to any attempt of modeling. All model constraints (mass balance, operational bounds for pipelines and processing units and product qualities, product demands, etc.) were then derived by the instructor after the students had enough time to try and formulate their propositions.

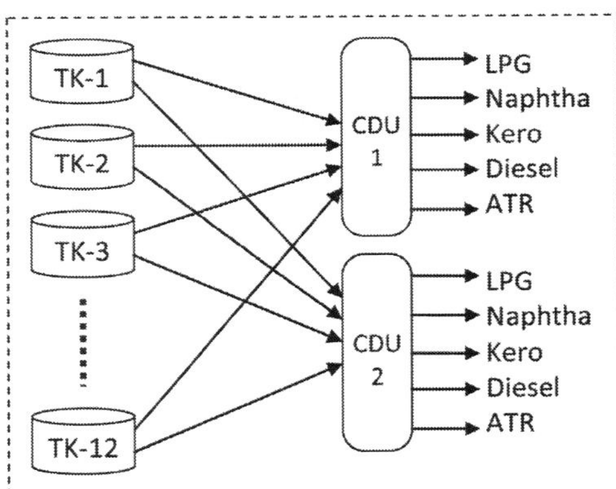

Figure 5: Schematic representation of exercise *xxxi*.

Lastly, the problem of crude oil unloading with inventory and crude blend quality management presented inLee et al. (1996) was addressed (Fig. 6). Two problem instances described in Lee et al. (1996) were modeled and solved: the *motivating example* (here "exercise *xxxii*") and *example 4* (here "exercise *xxxiii*"). The problem can in principle be modeled as a large-scale non-convex MINLP due to bilinear terms in property constraints. These non-convexities can be avoided by introducing individual entity flows. A rigorous MILP model derived from the previous MINLP is then obtained so as to ensure theoretical global optimality. Such linearization procedures (also illustrated in Joly and Pinto, 2003) cause an increase in the model size; nevertheless, it has the advantage of providing a better lower bound to the objective function. The most complex instance solved comprises 3 crude vessels, 6 storage tanks, 4 charging tanks, and 3 CDUs, which adds up to 3232 single variables, from which 540 are binary (0–1), and 4837 equations. Proven optimality was achieved after 29.8 CPU s (Intel Core i5 3.2 GHz 3.11 GB RAM DDR).

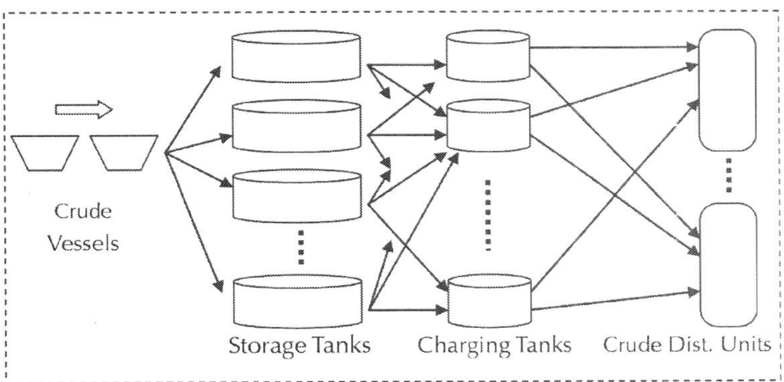

Figure 6: General schematic representation of examples *xxxii* and *xxxiii*.

Overview of Module 5

The aim of this module was to provide the necessary information to the learners about how easy it is to develop a flexible (for implementing customizations and linking to databases) and user-friendly (i.e., familiar) 'interface' for a GAMS application in order to automate data input/output operations, and enable refinery personnel without expertise on GAMS (and MIP) to run the optimization solution (Fig. 7).

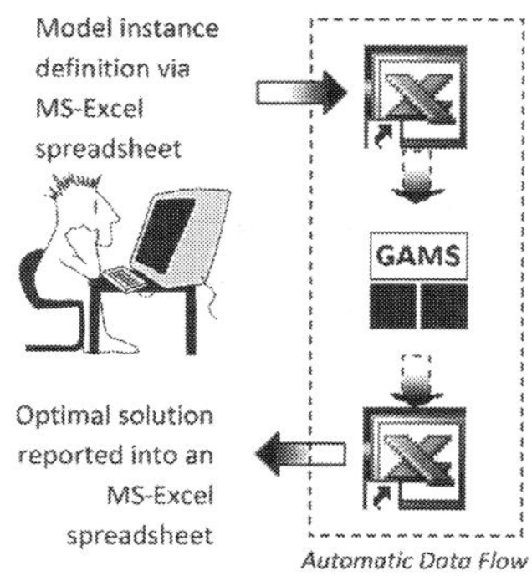

Figure 7: The integrated solution framework developed by the learners for the crude-oil scheduling problem from Lee et al. (1996) (exercises*xxxii* and *xxxiii*).

In particular, given the popular use of electronic spreadsheets in the refinery production management at PETROBRAS, the integration between GAMS and MS-Excel was focused on through the consideration of GAMS Data eXchange (GDX) files (McCarl, 2010). GAMS can read or write something called a GDX file, which is a binary file that can contain information regarding sets, parameters, variables and equations. Among other usages, GDX files can be used to prepare data for a GAMS model, pass results of a GAMS model into different programs, and pass results into GAMS from different programs. By addressing the problem from Lee et al. (1996) (Module 4), an GUI through which the user can entirely configure the model instance topology (number of crude vessels, storage and charging tanks, and CDUs), scheduling horizon and problem initial conditions were developed step-by-step. The application is completely operated through the MS-Excel interface and the solution process can be accompanied in a DOS window, as illustrated in Fig. 8. Once concluded, the optimal solution (crude vessel schedule, projected inventory levels, tank farm operation schedule, etc.) is automatically reported in a set of MS-Excel sheets, as illustrated in Fig. 9. If the instance is infeasible (or any numeric problem occurs), a message (pop-up screen) informs this condition to the user.

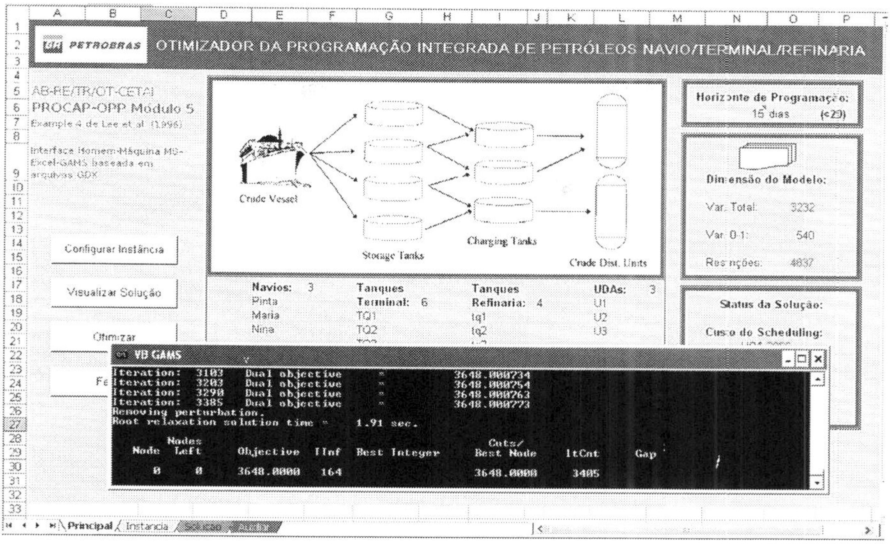

Figure 8: The application command display developed by the learners for the crude-oil scheduling problem from Lee et al. (1996) (exercises *xxxii* and *xxxiii*).

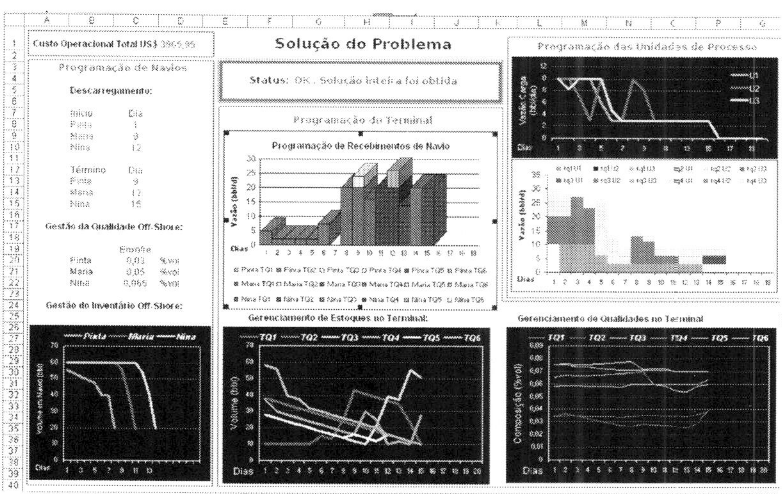

Figure 9: An example of automatic solution output display developed by learners for the crude-oil scheduling problem from Lee et al. (1996) (exercises *xxxii* and *xxxiii*).

Overview of Module 6

In this complementary module, the focus was to broaden the view of the learners regarding non-deterministic approaches to the optimization problem. As shown throughout this paper, a number of OR problems in the oil industry can be represented as sets of linear equations. This strategy not only makes the modeling task easier, but also fits the requirements of ordinary deterministic optimization algorithms.

However, the refining industry is plenty of non-linear relationships that arise from the non-ideal behavior of substances and production processes. It is hard not to imagine a future in which planning/scheduling problems will become more realistic, sharing the very same models like other automation layers, such as real time optimization (RTO) (Mendoza et al., 2013 and Feital et al., 2013) and model predictive control (MPC) applications (Moro and Grossmann, 2013). In this scenario, non-deterministic approaches should be more valuable than nowadays standard algorithms.

Major shortcomings of conventional optimization algorithms considered in the foregoing sections were discussed at the beginning of the course: the need of a *good* initial condition; the difficulty in coping with nonlinear behavior, such as presented by discontinuous, multi-modal and non-differentiable functions in high-dimension spaces. In contrast to the deterministic search for extreme points, the concept of random-walk and Markov chains were shown as the foundation of non-deterministic optimization algorithms. The first algorithm discussed was the Metropolis-Hastings Algorithm (MHG) (Hastings, 1970). Next, other algorithms were presented, such as Simulated Annealing (Kirkpatrick et al., 1983), Genetic Algorithms (GA) (Mitchell, 1996), Particle Swarm Optimization (PSO) (Kennedy and Eberhart, 1995) and Firefly Algorithm (FA) (Yang, 2010), according to the description shown in Table 12.

Table 12: Module 6 content

Topic	**Scope of learning**	**Solved examples**[a]	**Proposed exercises**[a]
Motivation	Historical overview of major advances in IT and OR algorithms along the last half century. Classification of OR algorithms. Metaheuristics definition and overview. Strengthens and weaknesses of metaheuristic approaches. Typical engineering problems successfully solved with metaheuristic approaches. Examples of hybrid approaches.	–	–
Stochastic Trajectory Optimization (STO)	Basic concepts. Mathematical formalization. Algorithm parameterization tuning. Random Walk, Markov Chains, Metropolis-type algorithm.	–	xxxv
Simulated Annealing (SA)	Basic concepts. Mathematical formalization. Algorithm parameterization tuning.	i	–
Genetic Algorithms (GA)	Basic concepts. The generic evolutionary algorithm. Algorithm parameterization tuning.	xxxii	xxxvi
Particle Swarm Optimization (PSO)	Basic concepts. Mathematical formalization. Geometric interpretation. Discrete-continuous approach. Algorithm parameterization tuning. Hybridization.	xxxii	xxxvi
Firefly Algorithm (FA)	Basic concepts. Mathematical formalization. Comparison among methods (STO/SA/GA/PSO/FA).	–	xxxvi

[a]Solved examples and proposed exercise themes are summarized in Table 5, Table 11 and Table 13.

Besides other minor didactic examples, the students had to solve the exercises described in Table 13. Solved examples *i* and *xxxii* are revisited exercises, since they had already been studied in the foregoing modules through MIP. An important issue regarding meta-heuristic methods is algorithm parameterization, which was fully discussed in exercise *xxxii*. By solving these examples, the learners captured the main differences, strengths and weaknesses of the proposed deterministic and stochastic approaches, being able to better understand the usefulness of each type according to the intrinsic nature of their problems.

Table 13: Summary of exercises

#	a	Subject
xxxv	●	Optimization of multi-modal objective functions using MHA.
xxxvi	●	Parameterization and setup of multi-modal optimization problems using PSO, GA and FA.

[a]Level of relative complexity. *Green*: low complexity; *red*: high.

COURSE ASSESSMENT BY LEARNERS

To ensure the quality of its educational portfolio, PCU has a standard evaluation process which is implemented right after a course is concluded. In this standard assessment, the learners are asked to give a quantitative feedback to a number of questions concerning general aspects of the course. This includes*technical* (e.g., quality of prior information about the course, adequacy of program scope and goals to their professional needs, and course schedule and duration – Table 14, questions 1–3), *methodological* (e.g., instructor's didactics, teaching methodology and interpersonal relationship– Table 14, question 4) and*infrastructure issues* (e.g., media and didactic material, classroom facilities and accommodations – data not shown). Additional specific questions were considered in order to assess the learner's ability to immediately apply the newly learned skills according to the extent they feel the course has helped them in achieving the expected outcomes (Table 14, questions 5–8).

Table 14: Learners' feedback

#	Methodological assessment (right after the end of the course)	Score[b]	SD[c]
1	Have you been clearly instructed in advance about the course objectives and on how it could be useful to improve your work and professional career as a process engineer?	9.3	0.9
2	Was the proposed curriculum adequate to meet the desired objectives?	9.4	0.9
3	Was the course schedule and duration well thought-out?	8.0	1.0
4	Was the teaching methodology adequate?	9.0	1.0

ID	Technical assessment performed right after the end of the course	Score	SD[b]
5	Ability to identify technical or economic opportunities to apply OR tools in order to increase the quality of the decision-making in refinery production planning and scheduling.	8.7	1.0
6	Ability to apply knowledge of mathematics and engineering to design and to conduct MIP-based optimization studies on simple-to-middle complexity scheduling problems on a professional modeling platform.	8.8	1.3
7	Ability to analyze a mathematical optimal solution and to perform a quantitative analysis of its global benefits, and to apply conclusions to the real system.	8.0	2.2
8	Ability to work in multi-disciplinary teams, and to explain and to discuss thoughts and results to other engineers.	8.2	1.6

ID	Technical assessment performed 3 years later	Score	SD[b]
9	Did this course motivate you to engage in life-long learning on OR/optimization techniques in order to turn you into a specialist on computer aided decision-making technologies inside the company?	9.0	0.4

10	With respect to other PCU courses you have participated in the past 5 years, what is the relative importance that one should attribute to this course regarding the new possibilities that the use of decision-making technologies may confer to the modern process engineer?	9.3	0.8
11	Apart from production planning and scheduling problems, has this course enabled you to address general decision-making problems in process systems engineering to meet desired needs?	8.7	1.2
12	Did this course enhance your ability to use knowledge of contemporary issues, skills, and modern engineering tools to perform critical analyses of work processes aiming at promoting more and better use of optimization in your working environment?	8.2	1.7

[a]Based on a single cohort, which is related to the first edition of this course.

[b]Average scores, which range from 0 (objective not met/not satisfied) to 10 (objective fully met/very satisfied).

[c]Standard deviation.

To assess individual abilities and to avoid total reliance on group-work skills (Gomes et al., 2006), at the end of each MIP learning module, the learners were required to submit evidence related to their achievements through a classroom "end-of-course (EOC) exam", which consisted in solving a (typically) higher complexity problem (problem *xi* in module 1; problem *xxiii* in module 2; problem *xxvii* in module 3 and problems *xxxii* and *xxiii* in module 4). No specific EOC exams were applied to the complementary modules 5 (graphical user interface GAMS/Excel) and 6 (meta-heuristics). To be approved, the learner had to obtain a score greater or equal to 7 (on a 0–10 scale). Proper problem modeling (8 points at most) and correct GAMS code implementation (2 points at most) represented the criteria adopted for assessing learners individually with respect to modules 1–4. The total average score was 9.6 (standard deviation 0.4).

However, implementing OR in an industrial setting is a challenging task that requires interdisciplinary staff commitment, managerial support and dedicated resources to produce tangible benefits which, in general, will require some time to be proven. In this regard, an additional course assessment was performed three years after the conclusion of its first edition in January 2011 (Table 14, questions 9–12). This complementary evaluation round was carried out in terms of the Accreditation Board for Engineering and Technology 2000 (ABET, 1998) program outcomes (summarized in El-Naas, 2011). Basically, the ABET 2000 focus is on *what students learned* rather than on *what material is taught*.

Overall, the learners' responses (21 respondents) indicated that they were satisfied with the curriculum and the teaching approach. Despite the inherent complexity of the subject addressed, complementary feedback in the subjective form (as anonymous commentaries and, in some cases, interviews) revealed that this educational program was considered one of the most useful learning they have had in the PCU, since it added meaningful value to their professional careers as process engineers in a number of ways:

- By enabling them to better operate sophisticated optimization-based tools which support core business areas of the company;
- By enabling them to identify (and model) trade-offs in decisions which impact profitability;
- By enabling them to locally develop powerful optimization-based tools through the use of modern computational techniques. Getting "autonomy" to efficiently solve real-world decision-making problems locally in the refinery was emphasized by learners as a major strength of this educational program;
- By enabling them to perform a critical analysis about the adequacy/effectiveness of existing key performance indicators (KPIs);
- By encouraging them to change ancient cultures and paradigms concerning the way the refining business is managed (e.g., rethinking current working processes);
- By enabling (and motivating) them to establish/increase collaborations with the company's R&D and IT services areas, responsible for the in-house development of computer-aided decision-making technologies aimed to support the company's operation.

- In addition, this course has motivated some participants to develop further specialization at the academic level (e.g., M.Sc. and Ph.D. programs) aiming at assuming a consultant position in the future. As a result, significant contributions to PETROBRAS have been produced since then (Nishioka et al., 2012, Menezes et al., 2014a and Menezes et al., 2014b).

The learner's subjective feedbacks (as anonymous commentaries) were rather general than specific with respect to describing their learning difficulties and in correlating these difficulties with the proposed didactical methodology. While potentially attractive to provide fast, normalized responses concerning scholastic assessments in academic settings, standard evaluation procedures (e.g., preformatted questionnaires) may be unsuitable in part to provide a complete and reliable assessment of the teaching effectiveness when a complex subject such as OR is taught in industrial settings. It should be pointed out that learners (employees) may feel uncomfortable about registering criticism in paper, even under anonymity. In this sense, interview-based assessments were more effective to properly discuss and to track didactical and methodological imperfections or *opportunities*.

As important outcomes of the assessment process, curricular and methodological changes were proposed for the next edition of this course, which will take place in 2015. A new module (24-h workload) was structured to show how the use of a commercial modeling platform specifically designed to solve OR/optimization problems in the chemical process industries can save of refinery planners and schedulers time to quickly prototype an optimization solution. The chosen platform was the IMPL by Industrial Algorithms LLC. Implementation of mentoring is another major novelty to be considered in the next edition of the course in the light of the learners' feedback. Instructors will support learners in modeling and solving a real-world refinery problem of their interest along the course. Here, the objectives are three-fold. Firstly, by studying a real-world problem of his/her particular interest, the learner commitment with the educational program may potentially increase. Secondly, our expectation is that mentoring can contribute to avoiding a period of total inactivity of the learner between two consecutive modules (note that there is typically a time gap of about 25–30 days between two adjacent modules). Thirdly, depending on the individual interest of the learner (and on the complexity of the problem addressed by him), the mentoring can evolve to a graduate program to be carried out in the academy under the support of Petrobras' advisors. Furthermore, if the problem addressed in the mentoring by a given learner can be generalized to other Petrobras refineries, the mentoring can also potentially evolve to the in-house development of a new, corporative OR solution. Such valuable opportunities corroborate the development of a new assessment policy which includes structured post-examination interviews of learners.

DISCUSSION

In this paper, we described a strategic, long-term educational program created by PETROBRAS in 2010 aimed to provide a practical, nevertheless comprehensive, view of OR foundations and large-scale optimization techniques for an interdisciplinary technical staff mainly composed by chemical (process) engineers. The educational literature on OR for chemical engineers remains scarce (e.g., García et al., 2012) and few works, if any, have been devoted to reporting educational experiences on teaching OR to chemical engineers working in the industrial environment. In this sense, the proposed curriculum, which was fully designed and implemented inside the company, may represent a primary reference for other chemical companies when discussing a similar educational agenda.

By focusing on key aspects of modern large-scale optimization algorithms relevant to solve (customized) decision-making problems in oil refineries, we stressed, on the basis of realistic examples of the oil industry, how to properly explore business opportunities (e.g., giveaway minimization) and to assess the corresponding economic benefit through using a professional modeling platform (e.g., GAMS). Good modeling techniques which are particularly relevant for modeling refinery planning and scheduling problems (e.g., quality constraints (Joly and Pinto, 2003 and Lee et al., 1996)) were taught. Realistic, nevertheless 'solvable' (i.e., low complexity), problems of great relevance for refinery planners and schedulers (e.g., in-line blending operations, crude oil area scheduling) were addressed along the course, which also taught the learners on how to develop a (complex) graphical user interface (GUI) for the developed MIP model, thereby rendering the application usable by other engineers without expertise on MIP techniques or the GAMS language.

Chemical engineers daily involved with complex decision-making processes related to refinery production planning and scheduling and supply chain logistics were the target public of this course. Rather than committed with theoretical aims, it was idealized and designed with the main purpose of nucleating and consolidating a new culture inside key operational areas of PETROBRAS. It should be pointed out that few, if any, chemical engineering undergraduate courses in Brazil include disciplines devoted to teaching OR or large-scale optimization techniques (e.g., MIP) in their standard scholar curricula. In Brazil, this valuable knowledge is normally acquired by chemical engineers solely via complementary formation (e.g., scientific initiation programs during the undergraduate course, M.Sc. and Ph.D. programs). As a result, few chemical engineers working in the industry are OR-experts in Brazil. We speculate that this may be the case in other countries as well, since local factors are important determinants for the best chemical engineering curriculum structure as well as for the nature and depth of specializations (Byrne, 2006).

From a technological point of view, the main contributions of this course are two-fold. Firstly, by acquiring theoretical background on mathematical modeling and optimization, the learners become able not only to make a better use of existing optimization tools in the refinery (e.g., by technically interpreting the optimal solution and exploiting sensitivity analysis), but also gained autonomy to work locally in the refinery (e.g., by performing model maintenance or developing new solutions). Secondly, since it democratizes the OR knowledge at the operational level, this course has significantly promoted collaborative work between R&D offices and operational areas of the company with regard to the in-house development of OR-based technology. This is a critical issue for keeping a sustainable competitive differential in this area. Whereas the in-house development of OR-based automation may be attractive in *strategic* (by preserving competitive know-how, enabling system integration and ensuring technological independence from the marketplace) and *tactical* terms (by allowing solution customizations), it may be very costly (typical investments may range from $ 250 K to $ 500 K per year per system). Furthermore, and principally, it requires a permanent, dedicated, highly qualified and multidisciplinary development team with a deep-level knowledge of supply chain operation, process engineering, OR and IT. In this field, providing the perfect understanding of the problem is the first (and the critical) step to successfully solve it. Planners and schedulers have detailed information about the problems they are concerned with and, hence, by qualifying these personnel on OR techniques (e.g., MIP) one can not only technically enable them to develop specialized, simple, OR solutions locally in the refinery, but also support the needs of a professional OR development team in a straightforward way.

As properly emphasized by Katajavuori et al. (2006), the most important goals of education should be to create learners who know how to learn and to foster the development of strategic, self-regulated, and reflective learners. Given the peculiar impact that OR technologies may have on the integrated performance of the refining business as well as on the whole company results, along this course, instructors' emphases were placed on how to stimulate learners to constantly reflect on and evaluate their learning experiences in the working environment. As a result, the major contributions of this course lied on the non-technological plan. Conversely to prior views coined in the early 1990s, when the massive introduction of personal computers and commercial refinery optimization tools (e.g., refinery planning systems) running locally at the refinery promised (and helped to achieve) new profitability thresholds, now it is clear that even more than any computerized systems, the quality of the staff is the prime success factor, along with their ability to work as a single team (Zhang and Valleur, 2010). In this sense, benefits were not limited to the technological plan. Since the end of its first edition in January 2011, this course has also encouraged refinery engineers to amplify their horizontal and vertical collaboration with other company's business areas concerned with supply chain optimization studies and performance monitoring. Most importantly, this educational program technically enabled and motivated the learners to: (a) rethink and to change inefficient work processes (e.g., those based on subjective decision-making, manual spreadsheets, phone and e-mails instead of integrated corporate IT and OR-based solutions based on reliable and standardized data), (b) question ancient managerial premises (e.g., *"Maximal processing charges all the time in my refinery!"*), and (c) break operational paradigms (e.g., mechanical (i.e., not optimal) selection and sequencing decisions, static blending recipes, constant set-points, etc.).

However, introducing a good educational program in the industrial environment may importantly differ from implementing a well-taught course in academic settings in a number of aspects. Hence, potential troublesome aspects should be identified in advance such that the learning environment can properly tackle those issues through specialized or even innovative ways. Here, at least four aspects had to be considered:

- *The human factor*. We were confronted with the learners' heterogeneity with respect to age, prior background, computational skills as well as specificities of their current and past professional experiences, professional expectations and aims. In this sense, tutoring (Wetzel, 1996) was adopted and provided multiple benefits to tutored learners, including increase in awareness and motivation to overcome individual-specific barriers;
- *The curriculum*. We had to deal with the fact that an educational (or more generally *a training*) program structured in the industrial environment is usually thought/designed to solve (very) specific corporative needs without a primary

concern about providing a comprehensive view of theory (a must in academic settings). Since learners arrived to the course with educational demands clearly well defined ("I have a problem, I need to learn how to solve it"), not only theory and practice had to be well weighted, but industrial examples that mirror their reality had to be addressed along the course. Once the usefulness of the (complex) curriculum proposed was demonstrated, the learner engagement was verified;

- *The goal.* Rather than oriented to enable the learner "to develop", an industrial training program on "refinery automation" technologies for operational staffs is usually devoted to qualifying the learner on how "to use" well a finished (typically a commercial) software solution. Refinery planning and scheduling may be considered a typical case. Aiming at going beyond what has been attempted in the area, our experience demonstrated that MIP can successfully be considered to introduce OR and large-scale optimization to refinery planners and schedulers;
- *The context.* We were faced with the challenging task of teaching the foundations of a complex science to highly demanded staffs. Refinery planning and scheduling personnel not only play a major role in the day-to-day operations but they are quite often regrouped with economic, logistic and supply chain studies, performance monitoring tasks and other high-profile activities. Hence, they cannot be out from their working offices for more than few days a month. In PETROBRAS refineries, this problem is particularly aggravated since the scheduling staffs are historically reduced from a head count viewpoint (1–3 schedulers against 4–10 schedulers in European refineries, for instance (Zhang and Valleur, 2010), mostly depending upon the complexity of refinery operation). Here, the sequential modular (LP → MILP → NLP → MINLP) curriculum along which theory and practice are intermittently merged proved to optimize the learning curve, thereby circumventing this drawback. In this

sense, it can be considered a candidate educational solution particularly for those chemical companies presenting similar educational needs and highly demanded staffs working in core business areas.

One potential limitation of the proposed course is that it neglects stochasticity. Deterministic models, while typically advantageous by allowing fast computation, can usually capture only the average behaviors. They cannot address questions about noise, measurement errors, and uncertainties of operational, logistic or commercial nature which may seriously impact refinery operation. Such topics are properly investigated with stochastic approaches.

While stochasticity is sometimes thought of as leading to small deviations from the deterministic prediction, it indeed may often lead to qualitatively different behavior. By looking at some of the addressed micro-systems (e.g., the blending unit, the crude oil area) one can show that, in addition to generally improving accuracy, stochastic analysis may in fact suggest fundamentally different resultant behavior patterns as compared to what the well-known deterministic treatment would predict. This in turn may have a significant impact on the overall operation of the macro-system (e.g., the entire refinery) in which these micro-systems are imbedded in. Regarding the expressive variability of key aspects under consideration by planners and schedulers, stochastic optimization (Shapiro and Philpott, 2009 and Bayraksan and Morton, 2006) is a complementary module (24-h workload) in the next version of this course in 2015. Based on MIP and GAMS as modeling platforms, the learning scope proposed comprises: (a) introduction to stochastic optimization (including a structured review of probability and statistics) (Ermoliev and Wets, 1988), (b) LP-based programs with uncertainty (Dantzig, 1955, Luedtke et al., 2008 and Sen and Higle, 1999), (c) problems with probabilistic constraints and ways of incorporating risk into stochastic programs (Kazemi et al., 2010), (d) chance constrained programming (Charnes and Cooper, 1958), and (e) Sample Approximation Approach (SAA) techniques to deal with a large number of scenarios (Luedtke and Ahmed, 2008).

CONCLUSIONS

Representing a first step toward the popularization of the OR science inside the PETROBRAS areas which generally have a high profile in the organization, the implementation of this long-term learning program was not expensive and offered a valuable interdisciplinary experience to process engineers involved with complex decision-making processes in the oil refining arena.

In the words of Zhang and Valleur (2010), "dealing with OR technology in the refining business requires a highly qualified staff including bright process engineers with long experience of the refinery and open minds to find new solutions and test new tools". By educating these employees on OR possibilities and MIP techniques, this paradigm-breaking course technically enabled them not just to create better pro-forma operational plans (through the best use of already implemented optimization tools or by upgrading their decision-making arsenal) but also to answer strategic and tactical business questions timely and accurately. This may be of paramount importance when market opportunities are detected (or unexpected crisis cannot be avoided) because, in these situations, the economic impact on the refinery performance may range from few thousands to several million dollars. Most importantly, by aggregating added-value to the planner's and scheduler's activity, this course also contributed in a straightforward way to the company's efforts on refinery automation improvement. Included here are the somewhat intangible benefits of nucleating new cultures and attitudes in the industrial environment, which catalyzed the redesign of inefficient work processes. The ability of this work to be replicated to other chemical industries can make this work appeal to a broader audience.

ACKNOWLEDGMENTS

M. Joly is deeply grateful to Dr. Geraldo M.D. Santos (Refining Technology Division, PETROBRAS Headquarters) for encouraging and supporting this educational program, and to the PROCAP Organization Committee members (Dr. Brenno C. Menezes and Dr. Lincoln F.L. Moro (Refining Technology Division, PETROBRAS Headquarters), Dr. Cristiane S. Pereira and Dr. Marcelo M.F. Oliveira (Logistics Division, PETROBRAS R&D Center)) by their active contribution in structuring this course. We are also indebted to Dr. Luiz Fernando J. Bernardo (Logistics Division, PETROBRAS R&D Center) and Dr. Roberto Iachan (OR Division, PETROBRAS IT) by their managerial involvement in making this corporate educational program a reality. Finally, we also thank Marcelo K. Hassimotto (Planning & Logistics Solutions Division, PETROBRAS IT) by tutoring portions of the GAMS course. In addition, the authors are grateful to the two anonymous referees for their insightful comments and constructive suggestions which led to an improved version of the paper.

REFERENCES

1. ABET, 1998. Engineering Criteria 2000, Available from: http://www.abet.org/eac/EAC 99-00 Criteria.htm#EC2000 (accessed 04.04.14).
2. Agrawal, A., Balasubramanian, K., 2006. Consider adopting next-generation refinery scheduling. Hydrocarb. Process. (September), 65–68.
3. Aires, M.A.C., Ribas, P.C., Joly, M., 2011. PLANREF – The Petrobras refinery planning solution. In: Process Systems Engineering Workshop 2011, Angra dos Reis, Rio de Janeiro,
4. Brazil.
5. Barsamian, A., 2007. Gasoline and Diesel Blending: Technology, Operations, Economics. Technical report. Refinery Automation Institute, New Jersey, NJ.
6. Bayraksan, G., Morton, D.P., 2006. Assessing solution quality in stochastic programs. Math. Prog. 108, 495–514.

7. Bazaraa, M.S., Sherali, H.D., Shetty, C.M., 1993. Nonlinear Programming: Theory and Algorithms, 2nd ed. John Wiley & Sons, New York, NY.
8. Bellman, R., 1957. Dynamic Programming. Princeton University Press, Princeton, NJ.
9. Benders, J., 1962. Partitioning procedures for solving mixed variables programming problems. Numer. Math. 4, 238–252.
10. Biegler, L.T., Grossmann, I.E., 2004. Retrospective on optimization. Comput. Chem. Eng. 28, 1169–1192.
11. Bonner and Moore Management Science, 1979. RPMS (Refinery and Petrochemical Modeling System): A system description.
12. Bonner and Moore, Houston, TX.
13. Brooke, A., Kendrick, D., Meeraus, A., 1998. GAMS – A User's Guide. The Scientific Press, South San Francisco, CA.
14. Byrne, E.P., 2006. The role of specialization in the chemical engineering curriculum. Educ. Chem. Eng. 1, 3–15.
15. Castillo, P.A.C., Mahalec, V., Kelly, J.D., 2013. Inventory pinch algorithm for gasoline blend planning. AIChE J. 59 (10), 3748–3766.
16. Charnes, A., Cooper, W.W., 1958. Chance constrained programming. Manage. Sci. 6, 73–79.
17. Cooper, W.W., 2002. Abraham Charnes and W.W. Cooper (et al.): a brief history of a long collaboration in developing industrial uses of linear programming. Oper. Res. 50 (1),35–41.
18. Dantzig, G.B., 1955. Linear programming under uncertainty. Manage. Sci. 1, 197–206.
19. Dantzig, G.B., 1963. Linear Programming and Extensions.n Princeton University Press, Princeton, NJ.
20. De Souza, G., Odloak, D., Zanin, A.C., 2010. Real time optimization (RTO) with model predictive control (MPC). Comput. Chem. Eng. 34 (12), 1999–2006.

21. Drud, A.S., 1996. A System for Large Scale Nonlinear Optimization, Reference Manual for Conopt Subroutine Library. ARKI Cons. & Develop. A/S, Bagsvaerd, Denmark.
22. El-Naas, M.H., 2011. Teaching water desalination through active learning. Educ. Chem. Eng. 6, e97–e102.
23. Ermoliev, Y., Wets, R.J.-B., 1988. Numerical Techniques for Stochastic Optimization. Springer-Verlag, Berlin, Germany.
24. Feital, T., Lima, P., Pinto, J.C., Souza Jr., M.B., Xavier, G., Lima, M.J., Joly, M., 2013. Rethinking petroleum products certification. J. Petrol. Eng., http://dx.doi.org/10.1155/2013/594368, Article ID 594368.
25. Ferreira, E.C., Lima, R., Salcedo, R., 2004. Spreadsheets in chemical engineering education – a tool in process design and process integration. Int. J. Eng. Educ. 20 (6), 928–938.
26. Ferris, M.C., 1998. MATLAB and GAMS: Interfacing Optimization and Visualization Software. Mathematical Programming Technical Report 98-19. Computer Sciences Department, University of Wisconsin, Madison.
27. García, N., Ruiz-Femenia, R., Caballero, J.A., 2012. Teaching mathematical modeling software for multiobjective optimization in chemical engineering courses. Educ. Chem.
28. Eng. 7, e169–e180.
29. Garey, M.R., Johnson, D.S., 1979. Computers and Intractability – A
30. Guide to the Theory of NP-Completeness. W.H. Freeman and Company, New York, NY.
31. Gomes, V.G., Barton, G.W., Petrie, J.G., Romagnoli, J., Holt, P., Abbas, A., Cohen, B., Harris, A.T., Haynes, B.S., Langrish, A.G., Orellana, J., See, H.T., Valix, M., White, D., 2006. Chemical engineering curriculum renewal. Educ. Chem. Eng. 1, 116–125.
32. Gomory, R.E., 1960. An Algorithm for the Mixed Integer Problem. Technical Report RM-2597. The Rand Corporation.

33. Guerra, O.J., Rodriguez, A.U., Montagut, S.M., Duarte, L.A., Angarita, J.D., 2010. A solution strategy for large-scale nonlinear petroleum refinery planning models. In: AIChE
34. Annual Meeting 2010, Minneapolis, Paper 356d.
35. Hastings, W.K., 1970. Monte Carlo sampling methods using Markov chains and their applications. Biometrika 57 (1), 97–109.
36. Hillier, F.S., Lieberman, G.J., 2005. Introduction to Operations Research, 8th ed. The McGraw-Hill Inc., New York, NY.
37. Iachan, I., 2009. A Brazilian experience: 40 years using operations research at Petrobras. Int. Trans. Oper. Res. 16, 585–593.
38. IBM, 2012. IBM® ILOG® CPLEX® Optimization Studio, Version 12 Release 2 Information Center, Available from: http://pic.dhe.ibm.com/infocenter/cosinfoc/v12r2/index.jsp
39. (accessed 04.04.14).
40. Joly, M., 2012. Refinery production planning and scheduling: the refining core business. Braz. J. Chem. Eng. 29 (2), 371–384.
41. Joly, M., Odloak, D., 2013. Rescue therapy planning based on HIV genotyping testing. Chem. Eng. Sci. 93, 445–466.
42. Joly, M., Pinto, J.M., 2003. Mixed-integer programming techniques for the scheduling of fuel oil and asphalt production. Chem. Eng. Res. Des. 81, 427–447.
43. Joly, M., Pinto, J.M., 2006. Role of mathematical modeling on the optimal control of HIV-1 pathogenesis. AIChE J. 52 (3), 856–884.
44. Karmakar, N., 1984. A new polynomial-time algorithm for linear programming. Combinatorica 4 (4), 373–395. 18 education for chemical engineers 1 0 (2 0 1 5) 1–19
45. Katajavuori, N., Lindblom-Ylanne, S., Hirvonen, J., 2006. The significance of practical training in linking theoretical

studiesm with practice. High. Educ. 51, 439–464.

46. Kazemi, M., Nourelfath, M., Ait-Kadi, D., 2010. A multi-stage stochastic programming approach for production planning with uncertainty in the quality of raw materials and demand. Int. J. Prod. Res. 48 (16), 4701–4723.
47. Kelly, J.D., Forbes, J.F., 1998. Structured approach to storage allocation for improved process Controllability. AIChE J. 44 (8), 1832–1840.
48. Kelly, J.D., Mann, J.L., 2003. Crude oil blend schedulingoptimization: an application with multimillion dollar benefits. Hydrocarb. Proc. 2003 (June), 47–53.
49. Kelly, J.D., Zyngier, D., 2007. An improved MILP modeling of sequence-dependent switchovers for discrete-time scheduling problems. Ind. Eng. Chem. Res. 46, 4964–4973.
50. Kennedy, J., Eberhart, R., 1995. Particle swarm optimization. In: Proceedings of IEEE International Conference on Neural Networks IV, Perth, WA, pp. 1942–1948.
51. Kirkpatrick, S., Gelatt Jr., C.D., Vecchi, M.P., 1983. Optimization by simulated annealing. Science 220 (4598), 671–680.
52. Kondili, E., Pantelides, C.C., Sargent, R.W.H., 1993. A general algorithm for scheduling batch operations. Comput. Chem. Eng. 17 (2), 211–227.
53. Land, H.A., Doig, A.G., 1960. An automatic method of solving discrete programming problems. Econometrica 28 (3), 497–520.
54. Lee, H., Pinto, J.M., Grossmann, I.E., Park, S., 1996. Mixed-integer linear programming model for refinery short-term scheduling of crude oil unloading with inventory management. Ind. Eng. Chem. Res. 35, 1630–1996.
55. Luedtke, J., Ahmed, S., 2008. A sample approximation approach for optimization with probabilistic constraints. SIAM J. Optim. 19, 674–699.
56. Luedtke, J., Ahmed, S., Nemhauser, G., 2008. An integer

programming approach for linear programs with probabilistic constraints. Math. Progr. 122, 247–272.

57. Magalhães, M.V.O., (Ph.D. Thesis) 2004. Refinery scheduling. Imperial College, London, UK.
58. Magalhães, M.V.O., Moro, L.F.L., Smania, P., Hassimoto, M.K., Pinto, J.M., Abadia, G.J., 1998. SIPP – A solution for refinery scheduling. In: NPRA Computer Conference, CC-98-145, San Antonio, TX.
59. Magatão, L., Arruda, L.V.R., Neves-Jr, F., 2011. A combined CLP-MILP approach for scheduling commodities in a pipeline. J. Sched. 14 (1), 57–87.
60. Maravelias, C.T., Sung, C., 2009. Integration of production planning and scheduling: overview, challenges and opportunities. Comput. Chem. Eng. 33, 1919–1930.
61. McCarl, B.A., 2010. Expanded GAMS User Guide Version 23.3. Texas A&M University.
62. Mendoza, D.F., Palacio, L.M., Graciano, J.E.A., Riascos, C.A.M., Vianna Jr., A.S., Le Roux, G.A.C., 2013. Real-time optimization of an industrial-scale vapor recompression distillation process. Model validation and analysis. Ind. Eng. Chem. Res. 52, 5735–5746.
63. Menezes, B.C., Kelly, J.D., Grossmann, I.E., 2014a. Improved swing-cut modeling for planning and scheduling of oil refinery distillation units. Ind. Eng. Chem. Res. 52, 18324–18333.
64. Menezes, B.C., Moro, L.F.L., Lin, W.O., Medronho, R.A., Pessoa, F.L.P., 2014b. Nonlinear production planning of oil-refinery units for the future fuel market in Brazil: process design
65. scenario-based model. Ind. Eng. Chem. Res. 53, 4352–4365.
66. Micheletto, S.R., Carvalho, M.C.A., Pinto, J.M., 2008. Operational optimization of the utility system of an oil refinery. Comput. Chem. Eng. 32, 170–185.
67. Mingo, L.M., Angarita, J.L., et al., 2011. Academic software

for optimizing supply-side bids in a wholesale electricity market. Comput. Appl. Eng. Educ. 19 (3), 457–465.

68. Mitchell, M., 1996. An Introdution to Genetic Algorithms. Mass. MIT Press, Cambridge, MA.
69. Moro, L.F.L., 2003. Process technology in the petroleum refining industry – current situation and future trends. Comput. Chem. Eng. 27, 1303–1305.
70. Moro, L.F.L., 2009. Optimization in the petroleum refining industry – the virtual refinery. In: Alves, R.M.B., Nascimento, C.A.O., Biscaia Jr., E.C. (Eds.), 10th International Symposium on Process Systems Engineering (PSE 2009). Elsevier, pp. 41–46.
71. Moro, L.F.L., Grossmann, I.E., 2013. A mixed-integer model predictive control formulation for linear systems. Comput. Chem. Eng. 55 (8), 1–18.
72. Moro, L.F.L., Zanin, A.C.Z., Pinto, J.M., 1998. A planning model for refinery diesel production. Comput. Chem. Eng. 22, S1039–S1042.
73. Neiro, M.S.N., Pinto, J.M., 2004. A general modeling framework for the operational planning of petroleum supply chains. Comput. Chem. Eng. 28, 871–896.
74. Nemhauser, G.L., Wolsey, L.A., 1993. Integer and Combinatorial Optimization. John Wiley & Sons, New York, NY.
75. Nishioka, G.K., Joly, M., Le Roux, G.A.C., 2012. Scheduling of offshore wells activities in petroleum specific resources. In: EngOpt2012 – 3rd International Conference on Engineering
76. Optimization, Rio de Janeiro, Brazil, p. 64, E-papers.
77. Pantelides, C.C., 1994. Unified Frameworks for Optimal Process Planning and Scheduling. Foundations of Computer Aided Process Operations (FOCAPO II), Colorado.
78. Pinto, J.M., Grossmann, I.E., 1998. Assignment and sequencing models for the scheduling of processes systems. Ann. Oper. Res. 81, 433–466.

79. Pinto, J.M., Joly, M., Moro, L.F.L., 2000. Planning and scheduling models for refinery operations. Comput. Chem. Eng. 24, 2259–2276.
80. Raman, R., Grossmann, I.E., 1991. Relation between MILP modelling and logical inference for chemical process syntheses. Comput. Chem. Eng. 15 (2), 73–84.
81. Rejowski, R., Pinto, J.M., 2003. Scheduling of a multiproduct pipeline system. Comput. Chem. Eng. 27, 1229–1246.
82. Reklaitis, G.V., 1991. Perspectives on scheduling and planning of process operations. In: 4th International Symposium on
83. Process Systems Engineering, Quebec, Canada.
84. Rocha, R., Grossmann, I.E., Aragão, M.V.S.P., 2009. Petroleum allocation at PETROBRAS: mathematical model and a solution algorithm. Comput. Chem. Eng. 33, 2123–2133.
85. Sen, S., Higle, J.L., 1999. An introductory tutorial on stochastic linear programming models. Interfaces 29 (2), 33–61.
86. Shapiro, A., Philpott, A.B.A., 2009. Tutorial on Stochastic Programming, Available from:
87. http://stoprog.org/index.html?SPTutorial/SPTutorial.html (accessed 18.12.12).
88. Stefansson, H., Sigmarsdottir, S., Jensson, P., Shah, N., 2011. Discrete and continuous time representations and mathematical models for large production scheduling
89. problems: a case study from the pharmaceutical industry. Eur. J. Oper. Res. 215, 383–392.
90. Steinschorn, D., Hofferl, F., 1997. Refinery scheduling using mixed integer LP and dynamic recursion. In: NPRA Computer Conference, Louisiana.
91. Symonds, G.H., 1955. Linear Programming: The Solution of Refinery Problems. Technical Report. Esso Standard Oil Co.
92. Swensen, M., Acuff, C., 2011. Rethink planning and scheduling's role in refinery optimization. Hydrocarb. Process. (June), 55–58, Special Report.
93. Trierwiler, D., 2005. Planning & Scheduling Decision

Technology Seminar. Technical report. Haverly Systems Inc., Houston, TX.

94. Tucker, M.A., 2001. LP modeling – past, present and future. In: NPRA 2001 Computer Conference, Dallas, TX.
95. Tynjälä, P., Valimaa, J., Sarja, A., 2003. Pedagogical perspectives on the relationships between higher education and working life. High. Educ. 46, 147–166. US Department of Defense, 1975. The Significance of Practical Training in Linking
96. Theoretical Studies with Practice. U.S. Department of Defense Training Document, Pamphlet, pp. 350–430.
97. Vavasis, S., 1991. Nonlinear Optimization: Complexity Issues. Oxford University Press, New York, NY. education for chemical engineers 1 0 (2 0 1 5) 1–19 19
98. Viswanathan, J., Grossmann, I.E., 1990. A combined penalty function and outer approximation method for MINLP optimization. Comput. Chem. Eng. 14 (7), 769–782.
99. Wetzel, M., 1996. Developing the role of the tutor/facilitator. Postgrad. Med. J. 72, 474–477.
100. Yang, X.S., 2010. Engineering Optimization. John Wiley and Sons Inc., Hoboken, NJ.
101. Zhang, N., Valleur, M., 2010. Refinery planning and scheduling. In: Handbook of Petroleum Refining and Natural Gas Processing. ASTM International, Conshohocken, PA (Chapter 18).

Chapter 3

Biotechnological Processes for the Refining of Petroleum

Sylvie Le Borgne and Rodolfo Quintero

Programa de Biotecnologı´a del Petro´leo, Instituto Mexicano del Petro´leo, Eje Central La´zaro Ca´rdenas No. 152, Col. San Bartolo Atepehuacan, 07730 Me´xico, D.F., Mexico.

ABSTRACT

This paper presents a general overview of petroleum biorefining or petroleum refining using biotechnology. An explanation is offered as to why modern biotechnology, due to important advances that have occurred in genetic engineering, the study of extremophilic microorganisms, protein engineering and biocatalysis in nonaqueous media, can be a good alternative to develop new biocatalysts to be used in the refining of petroleum. An important refining objective is the desulfurization of fuels, and recent advances in research and development on fuel biodesulfurization (BDS) are discussed in this paper. Other possibilities in petroleum biorefining exist for the denitrogenation of fuels, removal of heavy metals, transformation of heavy crudes into light crudes as well as depolymerization of asphaltenes, and are also being addressed in this paper. All these alternatives are currently realized at the level of research and development and there are no industrial applications as yet. The development of biocatalysts for biorefining processes will depend on the involvement of experts in many areas, ranging from biotechnology to petroleum refining.

INTRODUCTION

At the present time, petroleum refining is mainly based on the use of physicochemical processes such as distillation and chemical catalysis that operate under drastic conditions, i.e. high temperatures and pressures. Such processes are energetically costly and highly contaminant.

Biotechnology refers to the use of living organisms (or part of them) to make or modify products, improve plants or animals or develop microorganisms for specific uses. Biological catalysts, or biocatalysts, operate in a wide range of conditions, including ambient temperature and pressure and are endowed with a high selectivity, resulting in decreased energetic costs, low emissions and no generation of undesirable by-products. Whole microorganisms as well as their enzymes can use a wide range of compounds as substrates to transform them into specific products. Because of these unique characteristics, biotechnology is thought to be an interesting alternative or complement for the development of new petroleum refining processes. In this paper, we will use the word "biorefining" to refer to petroleum refining using biotechnology. Until now, the major applications of biotechnology in the petroleum industry have been limited to the microbiologically enhanced oil recovery

(MEOR), the production of single-cell protein, the treatment of waste streams (waste waters or gases) and the bioremediation of soils contaminated during the recovery, processing and distribution of petroleum [1]. There are no reports concerning the application of biorefining processes at the industrial scale yet. Until now, most of the work realized in petroleum biotechnology has generally been related to effluent treatment and bioremediation, implying that biorefining is as yet only at its infancy [2]. In the present paper, the authors will attempt to explain as to why it is reasonable to assume that modern biotechnology is expected to have a potential use in petroleum refining. For this, the authors offer a brief overview of the scientific knowledge on which this assumption is based. This is followed by a discussion of the possible areas of applications of biorefining.

BASIC SCIENTIFIC KNOWLEDGE SUPPORTING THE CONCEPT OF "BIOREFINING"

To develop biorefining processes, it is necessary to have biocatalysts that are resistant and active under the operating conditions usually found in the petroleum industry such as complex hydrophobic media and high temperatures. Important advances have occurred in the biological scientific knowledge in the areas of genetic engineering, extremophilic microorganisms, protein engineering and biocatalysis in nonaqueous media, which support the idea that biotechnology can be integrated into petroleum refining.

Genetic Engineering

Genetic engineering consists in modifying in a determinate way the genetic material of microorganisms of industrial interest so that they acquire new or enhanced capabilities. For this, novel DNA sequences are created by artificially joining together strands of DNA from different organisms through the use of recombinant DNA technology.

The design of recombinant microorganisms for petroleum biorefining includes the construction of microorganisms:

- able to transform the different types of compounds present in petroleum,
- possessing higher activities compatible with the design of efficient and economically viable processes,
- capable to secrete biosurfactants to increase the bioavailability of hydrocarbons to be transformed,
- stable under process conditions (e.g. solvent-resistant microorganisms).

Protein Engineering

Protein engineering makes possible the creation of proteins, especially enzymes, with new or enhanced properties by direct manipulation of their genes [3]. Protein engineering uses recombinant DNA methodologies, polymerase chain reaction (PCR) to amplify DNA fragments in vitro, high-throughput screening as well as specialized software for the analysis of protein structures.

The design of enzymes for petroleum biorefining includes the creation of enzymes that are:

- soluble and stable in organic media,
- able to accommodate hydrophobic substrates such as hydrocarbons at their active sites,
- able to transform a wide range of hydrocarbons,
- highly active and stable, compatible with the design of efficient and economically viable processes.

Biocatalysis in Nonaqueous Media

The advances realized in biocatalysis in nonaqueous media since 1984 opened new application fields to biotechnology [4]. Before this time, the application of biocatalysts had been restricted to aqueous media that were thought to be the only ones able to support biocatalytic reactions. It was demonstrated that enzymes are active and stable in organic and gaseous phases, as well as in supercritical fluids [5], [6], [7] and [8]. These systems are advantageous for the development of industrial biotechnological processes in which the substrates are generally insoluble in water and have to be dissolved in organic solvents. In the special case of petroleum, the fuel itself can be the organic solvent, avoiding or minimizing the addition of water in the fuel. The use of reaction media with low water contents increases the solubility and the bioavailability of hydrophobic substrates. Potential biocatalysts which may be used in organic media principally include enzymes and, in some cases, whole cells. Enzymes require less water than microorganisms to be active in organic media and, theoretically, only one film of water covering their surface should be sufficient for catalysis. However, in many cases, it is advantageous to use whole cells as in the case of complex transformations requiring the action of several enzymes organized in complex metabolic pathways and the presence of cellular cofactors (electron-transporting molecules). The isolation of microorganisms that are able to survive in the presence of high concentrations of organic solvents is promising for the design and development of biotransformations in organic media [9].

Extremophilic Microorganisms

Some environments, considered as extremes by man and apparently unable to support life, are colonized by especially adapted microorganisms called extremophilic microorganisms [10]. These environments include ecosystems with high or low temperatures (psychrophiles: optimal growth temperature <20 °C; thermophiles and hyperthermophiles: optimal growth temperature >50 and 75 °C, respectively), high pressures (barophiles: optimal growth at greater than 1 atm), extreme pH (acidophiles and alkalophiles: optimal growth at pH <3 and >10, respectively) and high salt concentrations (halophiles: optimal growth in 2.0–5.2 M sodium chloride). Many of these environments are similar to those found in the petroleum industry; therefore, extremophilic microorganisms and their enzymes, also called extremozymes, are attractive catalysts for use in petroleum refining. The enzymes isolated from extremophilic microorganisms possess unique properties that can be used for industrial applications: they are extremely thermostable and generally resistant to organic solvents and extreme pH. Therefore, there is an increasing interest in isolating and studying these microorganisms. So far, the practical use of enzymes from extremophiles in the petroleum industry has only been mentioned for petroleum recovery in which extremophilic hemicellulases could be used to break down guar gum for oil well stimulation [11].

The genes encoding extremophilic enzymes can be transferred to mesophilic hosts like the well-known gram-negative bacterium Escherichia coli in which the enzymes can be produced in larger quantities (recombinant enzymes). The recombinant extremozymes produced in mesophilic microorganisms generally conserve their activity and unique properties [12]. However, it is expected that metabolic pathways rather than purified enzymes will be needed for the transformation of complex substrates in biorefining applications. Recent advances in genetic engineering techniques for extremophilic microorganisms suggest that complex metabolic pathways relevant to biorefining could be transferred into extremophiles. Several genetic systems for extremophiles are available [13], [14] and [15]. Understanding of the amino acid sequences allowing extremozymes to function at extremes of temperature, salinity, pH and pressure should be used to redesign conventional enzymes to work in extreme conditions. For this, the most common approach is to compare protein structures of mesophilic and extremophilic homologous enzymes and then engineer the extremophilic structural features into the mesophilic enzymes [16]. Only relatively minor changes in protein structure would be responsible for the special properties of extremophilic enzymes, and no consensus have been determined yet [17]. So to date, no general strategy exists for converting, via site-directed mutagenesis, a mesophilic enzyme into an extremozyme. Another approach could be to directly select more robust variants of mesophilic enzymes by introducing an interesting gene or set of genes into an extremophile which is then grown in extreme conditions to select for mutations in the protein of interest [18]. In the last decade, microorganisms that are able to grow at temperatures up to 110 °C were discovered. These microorganisms are excellent candidates for the operating conditions found in refineries. Moreover, at high temperatures, the bioavailability and solubility of hydrocarbons are higher. Furthermore, halophilic environments present low water activities, and enzymes from halophiles might work particularly well in low water activity reaction media such as organic solvents. New extremophilic microorganisms possessing activities relevant to biorefining have therefore to be screened, in parallel to genetic engineering strategies. To exploit their potential, cultivation of extremophiles to high cell densities will have to be achieved, and recent success in cultivating a few extremophiles is encouraging [19] and [20].

BIODESULFURIZATION OF FUELS

The combustion of petroleum-derived fuels leads to the atmospheric emission of sulfur oxides (SOx) that contribute to acid rain and air contamination by particulate matter. To control these emissions, strict regulations on the sulfur content of fuels have been recommended worldwide. The last proposal of the Environmental Protection Agency of the United States (EPA) is to reduce the sulfur content of diesel from 500 ppm (today's regulation) to 15 ppm for year 2006 [21]. In Mexico, half of the reserves are Maya-type heavy crude oils with a 3.3% sulfur content [22]. Because the production of this type of oil is to be increased in the near future, the proportion of heavy crudes with high sulfur contents will increase in the refineries' feedstocks. Therefore, Mexico, as well as other petroleum-producing countries, needs new refining processes to reach the standards on sulfur content in fuels. Up to 70% of the sulfur in petroleum is found as dibenzothiophene (DBT) and substituted DBTs (methylated DBTs and benzo-DBTs) which are particularly recalcitrant to hydrodesulfurization (HDS). To comply with the environmental regulations on sulfur using HDS, higher and higher temperatures and pressures will be required, leading to increased costs and atmospheric emissions. Therefore, new alternative or complementary technologies to HDS are needed to help the refiners to comply with sulfur regulations.

A possible approach is the biocatalytic desulfurization or biodesulfurization (BDS) of fuels. BDS with anaerobic bacteria was first proposed in the 1950s by ZoBell [23]. However, significant advances were only made at the beginning of the 1990s when the aerobic bacterium Rhodococcus erythropolis IGTS8, which is able to selectively remove the sulfur from DBT without degrading the carbon skeleton of this compound, was isolated [24]. Since then, considerable efforts were made in research and development in BDS due to the implementation of progressively stricter sulfur content regulations in diesel. The model molecules used in BDS studies are DBT and its derivatives due to their recalcitrance to the traditional HDS process. The energetic value of fuels treated with the IGTS8 strain is not affected because DBT is not degraded but only transformed into 2-hydroxybiphenyl (2-HBP) which is recycled to the organic phase formed by the fuel while the sulfur is eliminated in the form of sulfate in the aqueous phase containing the biocatalyst. The first isolated and patented strain in 1992, R. erythropolis IGTS8, is the base of the BDS process proposed by the North American company ENCHIRA Biotechnology Corporation (ENBC), before Energy Biosystems Corporation (EBC). In this process, the biocatalyst, made by whole cells in an aqueous solution, is contacted with the fuel to be treated and the desulfurization reaction occurs at the interface between the aqueous and organic phases. The metabolic pathway for desulfurization, the corresponding enzymes and genes have all been elucidated [25], [26] and [27]. The desulfurizing activity is repressed by sulfate- and sulfur-containing amino acids [28]. The genes involved in DBT desulfurization constitute the "Dsz" desulfurization system. The Dsz pathway has also been called 4S pathway because it involves four enzymatic steps. Four enzymes participate in the conversion of DBT to 2-HBP, and they were designated as DszA, DszB, DszC and DszD. The DszA and DszC enzymes are monooxygenases that require a cofactor ($FMNH_2$) to perform the catalysis. This cofactor is regenerated by a flavin reductase enzyme, DszD, a process which depends on another cofactor, NADH [29]. The participation of these cofactors and the complexity of the metabolic pathway oblige working with whole cells. DszB is a desulfinase which releases the sulfur atom from the sulfinate by the DszA enzyme. BDS reactions were realized in middle distillates using wild-type or genetically enhanced strains, and different levels of desulfurization were obtained depending on the reaction conditions(Table 1) [30], [31] and [32].

Table 1: Some examples of middle distillates BDS

	Case A [30]	Case B [31]	Case C [32]
Biocatalyst	R. erythropolis I-19	Rhodococcus sp. ECRD-1	Rhodococcus sp. ECRD-1
Fraction to be treated	partially HDS-treated middle distillate	middle distillate	extensively HDS-treated middle distillate
Initial sulfur concentration (before BDS)	1850 ppm	20,500 ppm	669 ppm
Final sulfur concentration (after BDS)	615 ppm	14,400 ppm	56 ppm
Biocatalyst concentration	30 g l^{-1}	NR*	NR*
Middle distillate/water ratio	1:9	1:1000	5:1000
Reaction time	24 h	7 days	7 days
Temperature	30 °C	25 °C	25 °C

*NR=Not Reported.

Genetic engineering had to be used to improve the IGTS8 strain, leading to a 200-fold increase in DBT desulfurizing activity allowing shorter residence times compatible with a commercial application [33]. For this, modifications in media composition and growth conditions as well as genetic manipulation of the IGTS8 strain were performed, and DBT desulfurization rates around 20 μmol min^{-1} g^{-1} dry cells were obtained [34]. Genetic manipulation consisted in increasing dsz genes copy number, eliminating the sulfate repression by promoter change and deleting the last gene of the metabolic pathway dszB. Deletion of this gene eliminated the slowest step of the Dsz pathway and allowed the accumulation of hydroxybiphenyl sulfinate, which is a more valuable product than sulfate because it can be recovered from the aqueous phase and used as surfactant [33] and [34].

Sterically hindered DBT derivatives rather than DBT are the main target molecules to desulfurize when extremely low sulfur levels are desired. Strains able to desulfurize substituted DBTs have been reported[35] and [36]. As for chemical catalysis, DBTs substituted in positions 4 and 6 are the most difficult to desulfurize due to steric hindrance problems [37]. A genetically modified derivative of the IGTS8 strain, containing multiple copies of the dsz genes, was reported to desulfurize C5-DBT derivatives 10 times slower than DBT [30]. Desulfurization rates of recalcitrant DBT derivatives over 20 μmol min^{-1} g^{-1} dry cells are needed [38]. Efforts are currently made to improve desulfurization rates of these compounds using directed evolution methods, including gene shuffling and chemostat enrichment [39] and [40]. A first approach consisted in generating libraries of evolved dszC genes by a new in vitro DNA recombination method called "Random Chimeragenesis on Transient Templates" (RACHITT) [39]. Simultaneous improvement of desulfurization rate and range of oxidized substrates were obtained by this method. This study focused on DszC because this enzyme catalyzes the first and limiting step of the Dsz desulfurization pathway in terms of rate of oxidation of substituted DBT. A second approach consisted in chemostat enrichment for the directed evolution of gain-of-function mutants [40]. For this, two-phase sulfur-limited chemostats were performed in the presence of organosulfur compounds that did not serve as source of sulfur for the wild-type IGTS8 strain. The gain-of-function phenotype was due to a mutation in codon 261 in the dszC gene. Recently, the Petroleum Energy Center (PEC) of Japan reported the isolation of new thermophilic bacterial strains working at 50–60 °C, identified as Paenibacillus, and able to selectively desulfurize DBT and DBT derivatives without degrading their hydrocarbon matrix [41] and [42]. These strains follow the same metabolic pathway as R. erythropolis IGTS8 to perform the desulfurization [43]. The corresponding enzymes presented a high homology with the Rhodococcus enzymes; however, they were active at higher temperatures (50–60 °C). A moderately thermophile Bacillus subtilis strain was also shown to desulfurize DBT, 4,6 dimethyl-DBT and 3,4 benzo-DBT through the selective cleavage of carbon sulfur bonds [44]. These two thermophilic bacteria, Paenibacillus and Bacillus, exhibited lower activity than the previously reported mesophilic desulfurizing Rhodococci. More recently, a moderately thermophilic Mycobacterium strain with high desulfurizing activity toward DBT and DBT derivatives was reported [45]. The hydrophobicity of Mycobacteriacell surfaces is advantageous for the efficient uptake of hydrophobic compounds like DBT and its derivatives: mass transfer limitations are reduced with these bacteria (see below). Thermophilic strains could be used to develop a crude oil BDS process at higher temperatures at which viscosity of the crude is lower. Besides, if BDS could be performed at higher temperatures, it would therefore be unnecessary to cool the straight run or HDS-treated oil fractions to ambient temperature, and BDS could be integrated more easily to a refining stream.

Most of the published data on BDS concern biocatalyst characterization and improvement by classical microbiological methods and genetic engineering. There are very few reports on BDS process designs and cost analysis. Mass production of biocatalyst with high desulfurizing activity is very important for the development of an efficient BDS process. Some reports describe the high cell density culture of desulfurizing bacteria [46], [47] and [48]. Chemically defined media containing acetic acid, glycerol or sucrose as carbon sources were used in these studies. Biocatalyst preparation involves the use of traditional fermentation technology and would probably be realized outside the refineries. Further research is recommended to achieve the high cell density and cheaper cultivation of microorganisms potentially useful for BDS and other biorefining applications [49]. Feedstock preparation for the BDS reaction itself involves temperature and pressure reductions, depending on the biocatalyst used. With extremophilic microorganisms, it would not be necessary to work at ambient temperature or pressure, reducing the costs and operations associated to temperature and pressure adjustment. The BDS process consists in contacting the biocatalyst with the fuel, followed by separation of the desulfurized product, recovery and recycling of the biocatalyst. Because BDS involves a multistep metabolic pathway, whole cells rather than isolated enzyme reactions are needed. BDS reactions are performed in the form of whole resting cells which are non-growing live cells that retain most of the enzyme activities of growing cells. With resting cells, reactions can be performed at higher cell concentrations than with growing cells. The reaction takes place in the presence of an aqueous phase containing the biocatalyst and an energy source to ensure the physiological supply of electrons. The research reported in the area of bioreactor design employed stirred tank reactors, airlift reactors and emulsion phase contactor with free cells, as well as fluidized bed reactor with immobilized cells [33] and [50]. Most of the work has been made with stirred tank reactors [33]. The emulsion phase contactor was designed to create very small water/fuel/biocatalyst droplets to minimize the introduction of water into the fuel; however, very low desulfurization rates were reported with this system [50]. The fluidized bed configuration could be advantageous because the biocatalyst is immobilized on a matrix allowing continuous operation and easy product separation; however, it was reported that the biocatalyst requires further development for this configuration to be competitive. The reaction time and catalyst longevity should be of 1 and 400 h, respectively, for a cost-effective BDS process [33]. Recycling of the biocatalyst is necessary in an industrial process, and BDS biocatalysts must have sufficient operational stability. Desulfurizing Rhodococci with improved stability both in aqueous and water/oil media were recently reported by Monot et al. [51]. Separation of the desulfurized fuel from the biocatalyst might not be easy. In the case of Rhodococcus biocatalysts, a very stable fine emulsion (droplet sizes between 2 and 50 μm) is formed between the fuel and the cells [52]. The hydrophobic nature of Rhodococcus microorganisms makes them adhere preferentially to the oil–water interface in water–oil systems. This property is advantageous in terms of desulfurization rates because substrates are directly transferred from the oil to the cell, avoiding mass transfer limitations at this stage. However, this property can make difficult the separation of the desulfurized fuel from the biocatalyst, especially at the high cellular densities. The difficulty of breaking such emulsions by centrifugation led to the invention of new separation schemes based on filters and, more recently, hydrocyclones [53] and [54]. Details of these bioprocesses have not been published, so costs are difficult to evaluate. Several companies and public organizations, among them the US Department of Energy (DOE) and the PEC of Japan, invested about $110 million in research and development on BDS processes [55]. EBC announced the construction and operation of two BDS pilot plants for diesel, one of 5 bpd in 1999 with Total Raffinage and one of 5000 bpd in year 2000 with PetroStar in a small Alaskan refinery [56]. Recently, the US DOE contracted a 3-year project to PetroStar to develop a BDS biocatalyst using genetic engineering and design a 5000-bpd BDS pilot plant with ENBC [57]. However, scale-up of this process has been reported to be problematic [58]. The PEC announced the operation of a BDS pilot plant of 1 bpd of light crude oil for year 2000 [59]. It is estimated that the BDS process should result in reduced CO_2 emissions and up to 70–80% less energy consumed per barrel compared to HDS [60]. The estimated capital costs for the BDS process should be two-thirds those of HDS, whereas the operation costs should be reduced by about 15% [33] and [60]. In spite of all these research efforts, there is no commercial BDS process as yet.

In 1997, a Japanese refining firm published a study to forecast the environment faced by the refining industry in Japan [61]. Other regions of the world are expected to face similar pressures, among them more drastic environmental requirements and the need to process heavier crudes with high sulfur contents. In this study, it is reported that BDS should be considered in the design of future refineries even if improvements in BDS still have to be made. In 1999, EBC proposed several options for the integration of BDS into the refineries for diesel production [33]. In that time, the target was 50 ppm of sulfur in diesel and two possibilities were envisaged: BDS downstream of HDS or BDS in place of HDS. BDS was found to be competitive with HDS only in the case of concomitant production of hydroxybiphenyl sulfinates, with a "truncated" 4S metabolic pathway and commercialization of these bio-derived surfactants or starting materials for the synthesis of other useful chemicals [33] and [62]. BDS alone was unable to achieve the 50-ppm target for sulfur. Used downstream of HDS, BDS could reach this target. The new 15-ppm target for sulfur in diesel is another challenge and, as mentioned above, new biocatalysts able to attack the most recalcitrant molecules have to be developed, and research is performed in this direction at the present time. In this case, BDS as a complement to deep HDS could be an interesting option.

The use of recombinant Pseudomonas desulfurizing and solvent-tolerant strains to desulfurize DBT sulfones coming from chemical oxidation followed by solvent separation has recently been proposed [63]. The chemical oxidation of fuel's DBTs followed by the solvent separation of the corresponding sulfones led to a 10% decrease in the fuel's energetic value, which was unacceptable. BDS of the concentrated sulfones allowed the recuperation of the corresponding 2-HBPs that could be reintroduced to the fuel. This is an example of how chemical and biotechnological processes could be combined to solve the sulfur problem.

As mentioned before, enzymes require less water than microorganisms to be stable and active in organic media. In the case of BDS with microorganisms, significant quantities of water have to be added to the fuel to allow the microorganisms to work. The values found in the literature generally indicate low oil-to-water volumetric ratios. Patel et al. [64] reported BDS of DBT in model solutions containing 90% (w/w) hexadecane. Desulfurizing Rhodococci recently described by Monot et al. [51] were able to work in two-phase water/oil medium or in emulsified medium with volumetric water-to-oil ratio between 20:80 and 80:20. Mixing and separation of oil/water emulsions is not an easy task. Ideally, the desulfurization should be performed in the absence of water or with a minimum proportion of water. The enzymatic oxidation of DBT to its sulfone was proposed as an alternative BDS method [65] and [66]. For this, the hemoenzyme chloroperoxidase from Caldariomyces fumago was used as a biocatalyst. The reactions were performed at ambient temperature and atmospheric pressure with DBT dissolved in an organic solvent or in diesel, diluted in 15% acetonitrile in an aqueous buffer and in the presence of hydrogen peroxide. As DBT sulfone is more polar than DBT, it precipitates or partitions into the aqueous phase. DBT sulfone has a higher boiling point than DBT, allowing its separation by distillation. In the above-referenced works, distillation was used and it was possible to reduce the sulfur content of a straight run diesel from 16000 to 2700 ppm. However, the proportion of water in this process is still high and should be reduced. Important research efforts are being made to improve chloroperoxidase stability in organic media and also to favor the partition of the hydrophobic substrate (DBT) to chloroperoxidase active site rather than to the solvent.

The majority of the work on BDS was performed for middle distillates fractions. These results may serve as background for the desulfurization of other streams. Recently, the US DOE announced that research also started on gasoline BDS [67]. In streams used to produce gasoline, benzothiophenes (BTs) and thiophenes (THs) are important sources of sulfur in addition to DBTs. Microorganisms able to selectively desulfurize BTs were isolated [68], [69], [70] and [71]. The BT desulfurization pathway of Gordonia sp. strain 213E presents some homology with the IGTS8 DBT 4S desulfurization pathway and was proposed to proceed as: BT→BT sulfoxide→BT sulfone (activation of the sulfur by oxidation to facilitate opening of the thiophenic ring)→(z)-2-(2′-hydroxyphenyl)ethen 1-sulfinate (opening of the thiophenic ring)→2-(2′-hydroxyphenyl)ethan-1-al+sulfate (desulfination) [68]. It is interesting to mention that strain 213E is unable to desulfurize DBT and strain IGTS8 is unable to desulfurize BT [68]. Another approach for the desulfurization of gasoline has been proposed by Atlas et al. [72]. It consisted in introducing an oxygen atom to the sulfur heteroatom of thiophenic compounds with an IGTS8-derived BDS catalyst having a truncated metabolic pathway ending at the oxidation step. Once it has oxidized the sulfur atom of DBT, the ring structure of this compound is destabilized and transformed into thiophenes or other sulfur compounds after reacting with a FCCU catalyst. A Nocardia asteroides strain, which is able to use BT as the sole source of sulfur, could transform TH and 3-methyl-TH into more polar compounds. The gene encoding this activity was subcloned into E. coli and found to be homologous to mammalian dimethylaniline monooxygenases, also named flavin-containing monooxygenases [73]. These enzymes catalyze the oxidation of nucleophilic nitrogen, sulfur and phosphorus atoms in a wide variety of organic compounds. Crude oil would be another target for sulfur removal, and the Japanese are active in this area although few results have been published on this subject [41], [42] and [43].

OTHER POSSIBLE APPLICATIONS

Biodenitrogenation

Like sulfur, nitrogen is a petroleum contaminant which contributes to acid rain and atmospheric contamination. It also interferes with refining processes, leading to equipment corrosion and catalyst poisoning. Denitrogenation of petroleum fractions might facilitate their desulfurization by HDS. The removal of natural polar compounds in petroleum fractions by adsorption processes was found to improve the efficiency of HDS [74]. For example, a light gas oil feedstock, which was denitrogenated to the extent of 60% or higher by an adsorption pretreatment, resulted in a light gas oil product with a sulfur content below 100 ppm (wt.) at the same HDS operating conditions that would have produced a 300 ppm (wt.) sulfur product for the same untreated light gas oil feed. Like sulfur, nitrogen is found in heterocycles such as quinolines and carbazoles. Basic nitrogen compounds like quinolines are the most poisonous for catalysts. Recently, the isolation of a Pseudomonas ayucida strain, named IGTN9m, capable in selectively removing nitrogen from quinoline without degrading the carbon skeleton of this molecule, has been reported [75]. It might be important to remove nitrogen without destroying the hydrocarbon skeleton of the nitrogen compounds to keep intact the energetic value of the fuel [76]. Up to 5% of the nitrogen in the petroleum was removed with this strain. Other microorganisms able to selectively remove nitrogen from other nitrogen compounds present in petroleum may be isolated, and the metabolic pathways responsible for the denitrogenation may be studied both at the biochemical and genetic levels to be able to genetically engineer these strains [76]. Quinoline and carbazole degradative pathways have already been reported and some of them characterized at the biochemical and genetic levels [76], [77], [78] and [79].

Demetallation

Crude oil contains metals in the form of salts, petroporphyrins and other complexes in the asphaltenes [80]. The first are easily removed during the crude oil desalting process in which they concentrate into the aqueous phase. The removal of metals trapped in petroporphyrins and complexes is more problematic because porphyrins are embedded in extremely complex asphaltenic structures. Heavy metals (Ni and V) poison refining catalysts and are liberated to the atmosphere in the form of highly toxic oxides during fuel combustion.C. fumago chloroperoxidase enzyme was used to remove metals contained in petroporphyrins and asphaltenes and reductions of 93% and 53% for Ni and Va were reported [81]. Chloroperoxidase catalyzed the destruction of porphyrinic rings by a chlorination reaction, leading to the removal of the metals coordinated in the porphyrinic rings. Later, the oxidation of petroporphyrins and crudes using cytochrome Creductases from Bacillus megaterium and Catharanthus roseuse was reported [82]. These hemoproteins presented the advantage of liberating metals by oxidation of porphyrinic rings instead of chlorination, avoiding the formation of chlorinated products which are undesirable due to environmental concerns.

Transformation of Heavy Crudes into Lighter Crudes

Recently, the conversion of heavy crudes into lighter crudes utilizing microorganisms has been reported. This is undoubtedly of great interest for the petroleum industry due to the progressive depletion of high-quality light crudes [83], [84], [85] and [86]. For this, different bacterial genera such as Thiobacillus, Achromobacter, Pseudomonas and Sulfolobus were used. The bacteria were previously adapted to resist high temperatures, pressures, salt concentrations and hydrocarbon concentrations. These bacteria are extremophiles. The resistant strains were then introduced into the crudes in the form of a very concentrated aqueous solution and incubated at temperatures from 50 to 65 °C. The effect of microorganisms on crude oil depended both on the microorganism and the type of crude. Generally, the treated crudes became lighter and contained about 24–40% less sulfur, nitrogen, oxygen and heavy metals. It is thought that the microorganisms specifically directed their action to the heteroatoms and organometallic structures that serve as attachment and initiation points for microbial activity. The reactions involved probably consisted in oxidations and degradation of higher hydrocarbons. Asphaltenes were probably depolymerized by rupture at active sites containing heteroatoms, allowing the liberation of small molecules previously trapped. The increase in the concentration of saturated chains from C_8 to C_{26} indicated the degradation of high molecular weight hydrocarbons, probably alkanes. The decrease in heteroatoms and metals indicates that the heterocycles had possibly been oxidized into more soluble molecules that migrated to the aqueous phase containing the microorganisms. The specific microorganisms performing these transformations are not known with certainty and there is no available data concerning the involved biochemical reactions. Thus, more research is needed to elucidate the metabolic pathways responsible for these transformations and the corresponding genes so as to be able to improve these strains by genetic engineering. Chemical markers were used for the preliminary economic and technical analysis of this technology [83]. In the case of sulfur removal, it was found that reaction time and amount of sulfur removed were the most critical parameters to optimize to improve the overall cost-efficiency of this technology.

Depolymerization of coal asphaltenes by fungi, possibly through oxidative attack by the extracellular radical generating enzymes, has been reported before [87]. Such reactions could probably be envisaged for upgrading of petroleum asphaltenes.

CONCLUDING REMARKS

Even if most of the efforts in research and development have been directed to BDS, other research projects have been initiated for removing other contaminants from petroleum such as nitrogen and heavy metals. Projects concerning the transformation of heavy crudes into light crudes have also been reported. Depolymerization of asphaltenes is also envisaged. However, the most advanced area is BDS for which pilot plants were announced, while the other areas of application in biorefining are still at the level of basic research. It appears that in the distant future, the use of biotechnology could also be extended to other areas of petroleum refining, making possible hydrocarbon cracking, isomerization, polymerization or alkylation by biological catalysts [88]. The introduction of biocatalysis in petrochemistry is also expected [88]. The success of biocatalysis will depend on the ability to enhance biocatalyst activity (i.e. rate and range of substrates) and stability under the conditions found in the petroleum refining industry using genetic engineering strategies.

In all cases, the development of commercial biorefining processes will depend on significant improvements in the cheap and abundant production of highly active and stable biocatalysts adapted to the extreme conditions encountered in petroleum refining. Important improvements will also have to be realized in the design of bioreactors and phase contact and separation systems. Until now, BDS is the most advanced field of biorefining but no commercial application has been announced yet. The results obtained for BDS may be generally applicable to other areas of biorefining. The interdisciplinary participation of experts in biotechnology, biochemistry, refining processes and engineering will be essential.

ACKNOWLEDGMENTS

The present study was supported by the D.00020 research project of the Mexican Institute of Petroleum.

REFERENCES

1. G. Hamer, N. Al-Awadhi, Acta Biotechnol. 20 (2000) 335.
2. S. Le Borgne, N. Rojas, R. Quintero (Eds.), Proceedings First International Conference on Petroleum Biotechnology: State of the Art and Perspectives, Instituto Mexicano del Petro´leo, Mexico, 2000.
3. C. Schmidt-Dannert, F.H. Arnold, Trends Biotechnol. 17 (1999) 135.
4. A. Zaks, A.M. Klibanov, Science 224 (1984) 1249.
5. A. Zaks, A.M. Klibanov, J. Biol. Chem. 263 (1988) 3194.
6. J.S. Dordick, Enzyme Microb. Technol. 11 (1989) 194.
7. E. Barzana, A.M. Klibanov, M. Karel, Appl. Biochem. Biotechnol. 15 (1987) 25.
8. S.V. Kamat, E.J. Beckman, A.J. Russell, Crit. Rev. Biotechnol. 15 (1995) 41.
9. J.A.M. De Bont, Trends Biotechnol. 16 (1998) 493.
10. A. Aguilar, FEMS Microbiol. Rev. 18 (1996) 89.
11. E. Pennisi, Science 276 (1997) 706.
12. M.W.W. Adams, R.M. Kelly, Chem. Eng. News 73 (1995) 32.
13. V. Mai, J. Wiegel, in: A.L. Demain, J.E. Davies (Eds.), Manual

of Industrial Microbiology and Biotechnology, American Society for Microbiology, Washington, DC, 1999, p. 511.
14. K.R. Sowers, H.J. Schreier, Trends Microbiol. 7 (1999) 212.
15. J. Wayne, S.-Y. Xu, Gene 195 (1997) 321.
16. M.J. Danson, D.W. Hough, Trends Microbiol. 6 (1998) 307.
17. M.W.W. Adams, R.M. Kelly, Trends Biotechnol. 16 (1998) 329.
18. J.M. Weber, S.P. Johnson, V. Vonstein, M.J. Casadaban, D.C. Demirjian, Biotechnology 13 (1995) 271.
19. K.D. Rinker, C.J. Han, M.W.W. Adams, R.M. Kelly, in: A.L. Demain, J.E. Davies (Eds.), Manual of Industrial Microbiology and Biotechnology, American Society for Microbiology, Washington, DC, 1999, p. 119.
20. M. Krahe, G. Antranikian, H. Märkl, FEMS Microbiol. Rev. 18 (1996) 271.
21. S. Fletcher, Oil Gas J. Online Story (May 17, 2000).
22. PEMEX, Anuario estadístico de PEMEX, Mexico, 1998.
23. C.E. ZoBell, Process for removing sulfur from petroleum hydrocarbons and apparatus, U.S. Patent No. 2,641,564 (1953).
24. J.J. Kilbane, Mutant microorganisms useful for cleavage of organic CUS bonds, U.S. Patent No. 5,104,801 (1992).
25. S.A. Denome, E.S. Olson, K. Young, Appl. Environ. Microbiol. 59 (1993) 2837.
26. S.A. Denome, C. Oldfield, L.J. Nash, K.D. Young, J. Bacteriol. 176 (1994) 6707.
27. K.A. Gray, O.S. Pogrebinsky, G.T. Mrachko, L. Xi, D.J. Monticello, C.H. Squires, Nat. Biotechnol. 14 (1996) 1705.
28. M.Z. Li, C.H. Squires, D.J. Monticello, J.D. Childs, J. Bacteriol. 178 (1996) 6409.
29. L. Xi, C.H. Squires, D.J. Monticello, J.D. Childs, Biochem. Biophys. Res. Commun. 230 (1997) 73.
30. B.R. Folsom, D.R. Schieche, P.M. DiGrazia, J. Werner, S.

Palmer, Appl. Environ. Microbiol. 65 (1999) 4967.

31. M.J. Grossman, M.K. Lee, R.C. Prince, K.K. Garrett, G.N. George, I.J. Pickering, Appl. Environ. Microbiol. 65 (1999) 181.
32. M.J. Grossman, M.K. Lee, R.C. Prince, V. Minak-Bernero, G.N. George, I.J. Pickering, Appl. Environ. Microbiol. 67 (2001) 1949.
33. M.A. Pacheco, E.A. Lange, P.T. Pienkos, L.Q. Yu, M.P. Rouse, Q. Lin, L.K. Linguist, 1999 Annual Meeting, National Petrochemical and Refiners Association, San Antonio, Texas, USA, 1999, paper AM-99-27.
34. B.R. Folsom, in: S. Le Borgne, N. Rojas, R. Quintero (Eds.), Proceedings First International Conference onPetroleum Biotechnology: State of the Art and Perspectives, Instituto Mexicano del Petro´leo, Mexico, 2000, p. 103.
35. M.K. Lee, J.D. Senius, M.J. Grossman, Appl. Environ. Microbiol. 61 (1995) 4362.
36. A. Darzins, G.T. Mrachko, Sphingomonas biodesulfurization catalyst, U.S. Patent No. 6,133,016 (2000).
37. T. Ohshiro, T. Hirata, Y. Izumi, FEMS Microbiol. Lett. 142 (1996) 65.
38. D.J. Monticello, Curr. Opin. Biotechnol. 11 (2000) 540.
39. W.M. Coco, W.E. Levinson, M.J. Crist, H.J. Hektor, A. Darzins, P.T. Pienkos, C.H. Squires, D.J. Monticello, Nat. Biotechnol. 19 (2001) 314.
40. J.J. Arensdorf, A.K. Loomis, P.M. DiGrazia, D.J. Monticello, P.T. Pienkos, Appl. Environ. Microbiol. 68 (2002) 691.
41. J. Konishi, Y. Ishii, T. Onaka, K. Okumura, M. Suzuki, Appl. Environ. Microbiol. 63 (1997) 3164.
42. J. Konishi, Y. Ishii, K. Okumura, M. Suzuki, High-temperature desulfurization by microorganisms, U.S. Patent No. 6,130,081 (2000).
43. Y. Ishii, J. Konishi, H. Okada, K. Hirasawa, T. Onaka, M.

Suzuki, Biochem. Biophys. Res. Commun. 270 (2000) 81.

44. K. Kirimura, T. Furuya, Y. Nishii, Y. Ishii, K. Kino, S. Usami, J. Biosci. Bioeng. 91 (2001) 262.
45. T. Furuya, K. Kirimura, K. Kino, S. Usami, FEMS Microbiol. Lett. 204 (2001) 129.
46. H. Honda, H. Sugiyama, I. Saito, T. Kobayashi, J. Ferment. Bioeng. 85 (1998) 334.
47. M. Kishimoto, M. Inui, T. Omasa, Y. Katakura, K.-I. Suga, K. Okumura, Biochem. Eng. J. 5 (2000) 143.
48. J.H. Chang, Y.J. Kim, B.H. Lee, K.-S. Cho, H.W. Ryu, Y.K. Chang, H.N. Chang, Biotechnol. Prog. 17 (2001) 876.
49. D. Riesenberg, R. Guthke, Appl. Microbiol. Biotechnol. 51 (1999) 422.
50. B.L. McFarland, D.J. Boron, W. Deever, J.A. Meyer, A.R. Johnson, R.M. Atlas, Crit. Rev. Microbiol. 24 (1998) 99.
51. F. Monot, S. Abbad-Andaloussi, M. Warzywoda, Biological culture containing Rhodococcus erythropolis and/or Rhodococcus rhodnii and process for desulfurization of petroleum fraction, U.S. Patent No. 6,337,204 (2002).
52. A.P. Borole, E.N. Kaufman, M.J. Grossman, V. Minak-Bernero, R. Bare, M.K. Lee, Biotechnol. Prog. 18 (2002) 88.
53. J.C.T. Chen, D.J. Monticello, Method for separating a petroleum emulsion, U.S. Patent No. 5,525,235 (1996).
54. L.Yu, T. Meyer, B. Folsom, Oil/water/biocatalyst three phase separation process, U.S. Patent No. 5,772,901 (1998).
55. D.J. Monticello, Chemtech 28 (1998) 38.
56. Enchira Biotechnology.
57. Petro Star to develop biological sulfur removal process, Oil Gas J. Online Story (August 17, 2000).
58. Energy BioSystems advances biodesulfurization research, Oil Gas J. Online Story (May 26, 2000).
59. G. Ondrey, Chem. Eng. 105 (1998) 17.
60. L. Linguist, M. Pacheco, Oil Gas J. 97 (1999) 45.

61. M. Inomata, K. Sato, Y. Yamada, H. Sasaki, Oil Gas J. 95, 56.
62. E.A. Lange, Q. Lin, Compositions comprising 2-(2-hydroxyphenyl)benzenesulfinate and alkyl-substituted derivatives thereof, U.S. Patent No. 6,303,562 (2001).
63. http://www.ornl.gov/ORNLReview/v33_3_00/emissions.htm.
64. S.B. Patel, J.J. Kilbane, D.A. Webster, J. Chem. Technol. Biotechnol. 69 (1997) 100.
65. M. Ayala, R. Tinoco, V. Hernandez, P. Bremauntz, R. Vazquez-Duhalt, Fuel Process. Technol. 57 (1998) 101.
66. M. Ayala, N.R. Robledo, A. Lopez-Munguia, R. Vazquez-Duhalt, Environ. Sci. Technol. 34 (2000) 2804.
67. http://www.oit.doe.gov/factsheets/petroleum/pdfs/gasbiopet.pdf.
68. S.C. Gilbert, J. Morton, S. Buchanan, C. Oldfield, A. McRoberts, Microbiology 144 (1998) 2545.
69. C. Oldfield, N.T. Wood, S.C. Gilbert, F.D. Murray, F.R. Faure, Antonie van Leeuwenhoek 74 (1998) 119.
70. S.B. Kim, R. Brown, C. Oldfield, S.C. Gilbert, M. Goodfellow, Int. J. Syst. Bacteriol. 49 (1999) 1845.
71. M. Kobayashi, T. Onaka, Y. Ishii, J. Konishi, M. Takaki, H. Okada, Y. Ohta, K. Koizumi, M. Suzuki, FEMS Microbiol. Lett. 187 (2000) 123.
72. R.M. Atlas, D.J. Boron, W.R. Deever, A.R. Johnson, B.L. McFarland, J.A. Meyer, Method for removing organic sulfur from heterocyclic sulfur-containing organic compounds, U.S. Patent No. H1,986 (2001).
73. Y. Wang, J.D. Childs, C.H. Squires, Gene involved in thiophene biotransformation from Nocardia asteroides KGB1, U.S. Patent No. 6,235,519 (2001).
74. W.-S. Min, K. Choi, S.-Y. Khang, D.-S. Min, J.-W. Ryu, K.-S. Yoo, J.-H. Kim, Method for manufacturing cleaner fuels, U.S. Patent No. 6,248,230 (2001).

75. J.J. Kilbane, R. Ranganathan, L. Cleveland, K.J. Kayser, C. Ribiero, M.M. Linhares, Appl. Environ. Microbiol. 66 (2000) 688.

76. M.J. Benedik, P.R. Gibbs, R.R. Riddle, R.C. Willson, Trends Biotechnol. 16 (1998) 390.

77. S.-I. Sato, N. Ouchiyama, T. Kimura, H. Nojiri, H. Yamane, T. Omori, Appl. Environ. Microbiol. 179 (1997) 4841.

78. J.M. Shepherd, G. Lloyd-Jones, Biochem. Biophys. Res. Commun. 247 (1998) 129.

79. J. Schneider, R.J. Grosser, K. Jayasimhulu, W. Xue, B. Kinkle, D. Warshawsky, Can. J. Microbiol. 46 (2000) 269.

80. T.F. Yen, in: O.C. Mullins, Sheu (Eds.), Structure and Dynamics of Asphaltenes, Plenum, New York, 1998, p. 1.

81. P.M. Fedorak, K.M. Semple, R. Vazquez-Duhalt, D.W.S. Westlake, Enzyme Microb. Technol. 15 (1993) 429.

82. G.-W. Xu, K.W. Mitchell, D.J. Monticello, Process for demetalizing a fossil fuel, U.S. Patent No. 5,624,844 (1997).

83. E.T. Premuzic, M.S. Lin, H. Lian, W.M. Zhou, J. Yablon, Fuel Process. Technol. 52 (1997) 207.

84. E.T. Premuzic, M.S. Lin, J. Pet. Sci. Eng. 22 (1999) 171.

85. E.T. Premuzic, M.S. Lin, M. Bohenek, W.M. Zhou, Energy Fuels 13 (1999) 297.

86. E.T. Premuzic, M.S. Lin, Biochemical upgrading of oils, U.S. Patent No. 5,858,766 (1999).

87. M. Hofrichter, F. Bublitz, W. Fritsche, Fuel Process. Technol. 52 (1997) 43.

88. J. Aburto, N. Rojas, R. Quintero, Trends Biotechnol. (submitted for publication).

Chapter 4

Thermodynamic Analysis and Performance Optimization of an ORC (Organic Rankine Cycle) System for Multi-strand Waste Heat Sources in Petroleum Refining Industry

Jian Song[a], Yan Li[a], Chun-wei Gu[a], and Li Zhang[b]

[a]Key Laboratory for Thermal Science and Power Engineering of Ministry of Education, Department of Thermal Engineering, Tsinghua University, Beijing, China

[b]Sinopec Engineering Incorporation, Chaoyang District, Beijing, China

ABSTRACT

Low-grade waste heat source accounts for a large part of the total industrial waste heat, which cannot be efficiently recovered. The ORC (Organic Rankine Cycle) system has been proved to be a promising solution for the utilization of low-grade heat sources. It is evident that there might be several waste heat sources distributing in different temperature levels in one industry unit, and the entire recovery system will be extremely large and complex if the different heat sources are utilized one by one through several independent ORC subsystems. This paper aims to design and optimize a comprehensive ORC system to recover multi-strand waste heat sources in Shijiazhuang Refining & Chemical Company in China, involving defining suitable working fluids and operating parameters. Thermal performance is a first priority criterion for the system, and system simplicity, technological feasibility and economic factors are considered during optimization. Four schemes of the recovery system are presented in continuous optimization progress. By comparison, the scheme of dual integrated subsystems with R141B as a working fluid is optimal. Further analysis is implemented from the view of economic factors and off-design conditions. The analytical method and optimization progress presented can be widely applied in similar multi-strand waste heat sources recovery.

INTRODUCTION

With the development of science and technology, the energy demand is enlarging rapidly, which results in growing consumption of primary fossil fuels and massive discharge of pollutants, especially in China, the biggest developing country in the world. The energy consumption in industry is about 70% of the total energy consumption in China and more than 50% of the industrial waste heat could not been effectively recovered [1]. With intense concerns for energy conservation and emission reduction, more and more attention has been paid on waste heat recovery.

The waste heat sources are generally categorized by temperature levels: low-grade (<230 °C), medium-grade (230–650 °C) and high-grade (>650 °C). Medium and high-grade waste heat sources are easy to be recovered through steam Rankine cycle and thermal storage technology [2]. However, low-grade heat scurce, which accounts for more than half of the total, is difficult to be utilized. The ORC (Organic Rankine Cycle) system has been proved to be a promising energy conversion technology for the utilization of low-grade heat sources due to its significant advantages such as robust thermal performance under low temperature conditions, system simplicity, and environmental and economic benefits as well. A lot of research has been done on ORC systems. Saleh et al. [3] made comparison of the thermal characteristics among 31 pure working fluids for one heat source, of which the temperature is between 100 °C and 30 °C. The results showed that the cycle efficiency of dry and isentropic fluids was higher than that of wet fluids. Roy et al. [4] studied parametric optimization and performance analysis of a waste heat recovery system based on ORC, using R-12, R-123 and R-134a as working fluids, and the results showed that system using R-123 had the maximum work output and highest efficiency. Sun et al. [5] studied the influence of controlled variables, including working fluid mass flow rate, air mass flow rate of the condenser, turbine inlet pressure and inlet temperature, on the system performance, with the aim of achieving the optimal operating variables for maximizing the net power output. Zhou et al. [6] presented an experimental recovery system, in which R-123 was selected as a working fluid, a scroll expander was used to produce work, and fin tubes heat exchanger was designed as an evaporator. Skarke et al. [7] made a feasibility analysis of the application of ORC as a waste heat recovery system for automotive internal combustion engines, and the results showed that a simple ORC system was able to recover up to 10% of the engine waste heat.

Much research work has been done in thermal design, system analysis and performance optimization of ORC systems. Most of the researchers focused on utilizing one single waste heat source other than multi-strand waste heat sources. However, it is clear that there might be several waste heat sources distributing in different temperature levels in one industry unit, and it can be easily understood that the entire recovery system will be extremely large and complex if the different heat sources are utilized one by one through several independent ORC systems, which may also lead to high capital input and large space occupation.

There are five waste heat sources in a 1.2 million ton-level reforming and extraction unit in Shijiazhuang Refining & Chemical Company, SRCC, in China, which distribute in different temperature levels. Based on the research [8] of single ORC system and the 500 kW ORC test system developed by Tsinghua University and Hangzhou Chinen Steam Turbine Power Co., Ltd., thermodynamic analysis and performance optimization are implemented for the recovery system utilizing these five waste heat sources in this paper. The maximum net power output is chosen as the evaluation criterion to select suitable working fluids and optimal set of thermodynamic parameters, while the system simplicity, technological feasibility and economic factors are considered in the optimization. Four schemes are presented in continuous optimization progress, and the simulation results indicate that the scheme of dual integrated subsystems with R141b as a working fluid is optimal and recommended for practical uses in this heat recovery case.

ORGANIC RANKINE CYCLE SYSTEM

Thermodynamic Model of the ORC System

Fig. 1 illustrates the system of a basic Organic Rankine Cycle system for waste heat recovery. The liquid organic working fluid from the condenser is pumped to the preheater, heated and then transported to the evaporator. In the evaporator, the liquid working fluid is heated and turns into saturated or superheated vapor. Hence, in the turbine, the organic vapor expands to a low pressure state, driving the turbine to produce power, which is turned into electricity by a generator connected to the shaft with the turbine. Afterwards, the exhaust organic gas from the turbine outlet is condensed to liquid again in the condenser by cooling water.

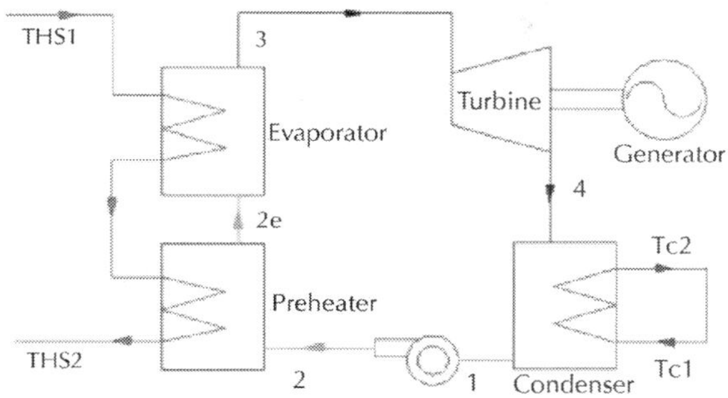

Figure 1: Schematic diagram of an Organic Rankine Cycle system for waste heat source recovery.

The process of Organic Rankine Cycle is shown in the T-s diagram in Fig. 2, which can also be described as follows:

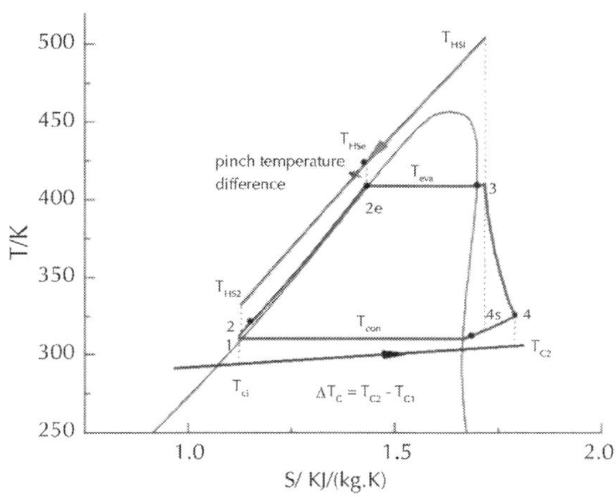

Figure 2: T–s diagram of Organic Rankine Cycle.

Mass flow rate of the organic working fluid

$$G_W = \frac{Q_{total} \cdot (1 - \varepsilon_Q)}{h_3 - h_2} \tag{1}$$

Where Q total is the heat load of waste heat source, and $_Q$ is the heat loss of the heat exchanger. As the beginning and ending temperatures of the waste heat source are fixed in this research, the heat load absorbed by ORC system can be calculated as

$$Q_{total} = G_S \cdot C_{P.S} \cdot (T_{HS1} - T_{HS2}) \tag{2}$$

Where $C_{P.S}$ is the average specific heat capacity of fluid or gas of the waste heat source, and T_{HS1} and T_{HS2} are defined as the beginning and ending temperature.

Process 1 to 2 (Pump)

$$W_P = \frac{G_W \cdot (h_{2rs} - h_1)}{\eta_{pump}} \tag{3}$$

Where h_{2rs} is the isentropic enthalpy of the working fluid after compressed in the pump, and $_{pump}$ is the efficiency of the pump.

Process 2 to 2e (Preheater)

$$Q_{prh} = \frac{G_W \cdot (h_{2e} - h_2)}{1 - \varepsilon_Q} \tag{4}$$

Process 2e to 3 (Evaporator)

$$Q_{eva} = \frac{G_W \cdot (h_3 - h_{2e})}{1 - \varepsilon_Q} \tag{5}$$

Process 3 to 4 (Turbine)

$$W_T = G_W \cdot (h_3 - h_{4s}) \cdot \eta_T \tag{6}$$

Where h_{4s} is the isentropic enthalpy of the exhaust gas in the turbine outlet, and $_T$ is the efficiency of the turbine.

Process 4 to 1 (Condenser)

$$G_C = \frac{G_W \cdot (h_4 - h_1)}{(1 - \varepsilon_Q) \cdot C_{P.C} \cdot (T_{C2} - T_{C1})} \tag{7}$$

Where $C_{P.C}$ is the average specific heat capacity of cooling water.

The net power output of the system is

$$W_{net} = W_T - W_P \tag{8}$$

Here the power required by auxiliaries is neglected since its amount is considerably small and the thermal efficiency is defined as

$$\eta_e = \frac{W_{net}}{Q_{total}} = \frac{W_T - W_P}{Q_{total}} \tag{9}$$

Based on the experiments conducted on the 500 kW ORC test system [8], some design parameters are given in Table 1.

Table 1: Design parameters of the ORC system

Property	**Symbol**	**Value**
Pump efficiency	ηpump	0.8
Heat loss of heat exchanger	εQ	0.01
Preheater pressure loss	ζQex	0.08
Minimum temperature difference of evaporator	Δteva/K	6.0
Turbine inlet pipe pressure loss	$ζP_1$	0.02
Turbine efficiency	ηT	0.8
Turbine outlet pipe pressure loss	$ζP_2$	0.015
Cooling water initial temperature	TC_1/K	293.15
Cooling water temperature rise	ΔTC/K	8.0
Minimum temperature difference of condenser	Δtcon/K	6.0

Working Fluid Selection

Generally, organic working fluids can be divided into three categories, dry, isentropic and wet, according to the saturation vapor curve slope, as shown in Fig. 3. A large amount of study has been conducted on selecting suitable working fluids for ORC systems under different conditions. Hung et al. [9] proved that the major physical properties affecting the system performance included latent heat, specific heat of liquid and vapor. Chen et al. [10] suggested that fluids with high latent heat and low liquid specific heat were preferable. However, Yamamoto et al. [11] suggested low latent heat fluid was better.

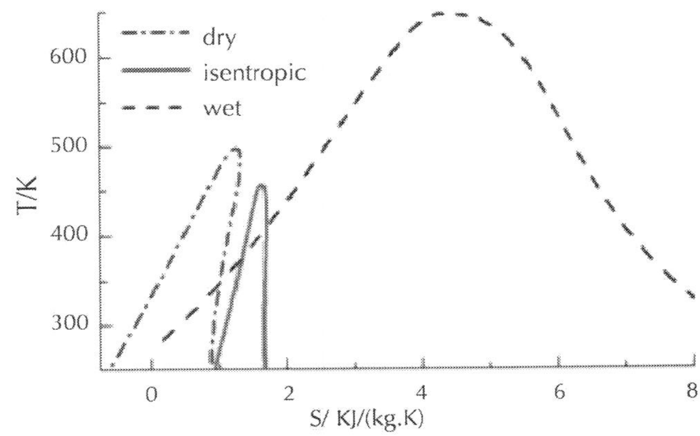

Figure 3: T–s diagram of dry, isentropic and wet organic fluids.

In this paper, the organic working fluid candidates are selected according to following principles:

- Proper thermal properties for the recovery system;
- Chemical stability under the operating condition;
- Slighter impact on the environment (low GWP (Global Warming Potential) and low ODP (Ozone Depletion Potential));
- Security during the cycle (non-flammable, non-explosive and non-toxic); and
- Availability and low cost.

BACKGROUND OF THE WASTE HEAT SOURCES

Petroleum industry is one of the key industries in China, which relies heavily on the primary energy resources. The heat load of the waste heat in petroleum industry is so vast that the recovery of such has great potentialities. Liu et al. [12] presented models and analysis of the energy-savings potential for refining and conversion processes in the context of technological change. The results indicated that upgrading process heaters had been a priority during recent years, but heat recovery and advanced process control systems would gradually begin to dominate the technological marketplace in the long term. Varga et al. [13]summarized the results of an engineering study of partial substitution of an air cooler, which was situated in one of the main process units of the refinery and cooled the process stream from 140 °C to 45 °C meantime dissipated heat of 32 MW into environment, with an organic Rankine cycle. Huang [14] made analysis of an ORC system using R601A as a working fluid, in order to utilize the low-grade waste heat in refining units. Meacher [15] designed a low temperature organic Rankine cycle system using R-113 as a working fluid for recovery and conversion of process waste heat for typical application in oil refineries and chemical plants.

However, in most running petroleum factories in China, the waste heat is simply carried away by cold water, which mainly refers to the water equilibrium and the heat equilibrium. Such kind of cooling method requires a large amount of cold water, while the reserve of water is low in most places of China. At the same time, such method doesn›t meet the requirement of energy saving. In recent years, the Chinese government has introduced new policies in order to control the enlarging consumption of fossil fuels and the massive discharge of pollutants. All the running petroleum factories are requested to be renovated and equipped with waste heat recovery systems, with the goal of cutting down the amount of consumption of water used as a cooling medium and utilizing the waste heat.

This paper focuses on the recovery of waste heat sources distributing in various temperature levels from the catalytic reforming unit in Shijiazhuang Refining & Chemical Company, SRCC, in Hebei Province, China, where the water resource is considerably lacking. The renovation of the unit is mainly conducted by Sinopec

Engineering Incorporation, SEI, and it is currently in the process of implementation. According to the temperature conditions of the waste heat sources, the Organic Rankine Cycle technology is selected with the consideration of cold water saving. The system design and performance optimization is conducted by Tsinghua University and SEI. The beginning temperatures of the waste heat sources are 215 °C, 205 °C, 147 °C, 104 °C and 98 °C, and the ending temperature is 80 °C as required by the technological process. The corresponding heat loads are 2.20 MW, 3.79 MW, 8.59 MW, 11.50 MW and 6.50 MW. The data of the five waste heat sources are listed in Table 2.

Table 2: Five waste heat sources to be recovered

Term	Source	Temperature level (°C)	Heat load (MW)	Proportion
A	Bottom oil of xylene	215–80	2.20	6.75%
B	Xylene products	205–80	3.79	11.63%
C	Evaporation tower	147–80	8.59	26.37%
D	Reforming tower	104–80	11.50	35.30%
E	Benzene tower	98–80	6.50	19.95%
Total			32.58	100%

THERMODYNAMIC ANALYSIS AND PERFORMANCE OPTIMIZATION

In this section, four schemes of the comprehensive ORC system for the five waste heat sources are presented in continuous optimization progress and comparison is made from the view of thermodynamics, system simplicity and technological feasibility. Since the ending temperatures of heat sources are set to 80 °C, the total heat loads of the heat sources are fixed. Thus the design of the system aims to achieve the maximum net power output, which represents the overall heat recovery efficiency. The condensing temperature is set to 34 °C according to the ambient condition.

Five Independent ORC Subsystems

It is easy to come up with the design of five independent ORC subsystems of different parameters to recover the five waste heat sources one by one. According to the principle of working fluid selection mentioned above, several organic fluids are selected as candidates. The properties of the working fluids are calculated by REFROP 9.0 [16] developed by the National Institute of Standards and Technology of the United States. The system simulation is carried out by a computer program written by authors under FORTRAN environment.

For each waste heat source and each kind of working fluid, the simulation program arrives to a proper evaporation temperature through continuous iteration, the heat load above which is utilized to convert the working fluid liquid into vapor in the evaporator and the heat load below which is used in the preheater. The computation results are listed in Table 3.

Table 3: Simulation results of the five independent subsystems

Heat sources	Working fluid	Evaporation tempera-ture, K	Mass flow rate, kg/s	Net power output, kW	Thermal efficien-cy, %
A	R123	429.7	9.63	350.08	15.93
	R601	428.2	4.02	337.26	15.35
	R601a	433.2	4.20	335.97	15.29
B	R123	412.2	16.93	558.81	14.74
	R601a	422.7	7.39	557.20	14.70
R245ca 420.2			13.86	556.28	14.68
C	R600a	381.7	21.43	938.27	10.92
	R236ea	381.5	43.42	930.75	10.83
	R245fa	375.0	37.04	923.43	10.75

D	R141b	352.2	45.21	961.30	8.36
	Heptane	352.9	25.97	952.42	8.29
	R601	353.4	26.35	948.50	8.25
E	R141b	350.9	25.63	530.43	8.16
	2-Butene	351.4	14.90	520.22	8.01
	R601a	351.7	14.99	519.49	7.99
Total	Maximum net power output and highest thermal efficiency			3338.89	10.25
	Hydrocarbons as working fluids			3300.73	10.13

As shown in Table 3, R123 is a working fluid suitable for waste heat sources A and B, because its critical temperature is relatively high, which is a little lower than the beginning temperatures of heat sources A and B, at 215 °C and 205 °C. R600a is a working fluid suitable for heat sources C for the same reason. The temperature of heat source D and E is low and the temperature drop is relatively small. Systems with R141b as a working fluid have a good performance for heat source D and E though the critical temperature of R141b is fairly high. The main reason is that the latent heat is in a larger amount than the sensible heat at a low evaporating temperature for the organic fluid whose critical temperature is high. Thus systems with these kinds of working fluids give out comparatively better thermal performance.

The scheme of five independent subsystems gives out a maximum net power output at 3338.89 kW and the overall thermal efficiency at 10.25% respectively. It is worth mentioning that hydrocarbons are products of the company and are easily available. Even though hydrocarbons are flammable and explosive, the cost possibly incurred from such related accidents are not considered in this paper, as the company is equipped with the explosion proof systems. The maximum total net power output of the system with hydrocarbons as working fluids reaches 3300.73 kW, slightly lower than the former. The scheme above is theoretically simple. But it is evident that the entire system including five independent subsystems is complex in structure, and requires high capital cost and large space occupation.

Single ORC System

The single ORC system for all five heat sources is the simplest in structure. Several heat exchangers are designed in this scheme to match the heat loads for both evaporation and preheating. As shown in Table 2, heat sources D and E account for more than half of the total heat loads while the high-grade parts, A and B, only make up 18%. If the evaporation temperature is set too high, the mass flow rate of working fluid will be low. Thus heat load needed for preheating is much smaller than the remaining part after evaporation, and the ending temperature of waste heat source will be higher than 80 °C. Conversely, a proper evaporation temperature may be low, resulting in high irreversible loss and low thermal efficiency of the system. The schematic diagram of this scheme is shown in Fig. 4 and the simulation results of the single ORC system are shown in Table 4, which shows that the thermal efficiency is quite low due to the low evaporation temperature. The maximum net power output of this scheme is just 2813.02 kW, nearly 16% lower than that of the scheme of five independent subsystems. Therefore, the single ORC system, in spite of the simplicity, is not efficient enough and neglected in further discussion.

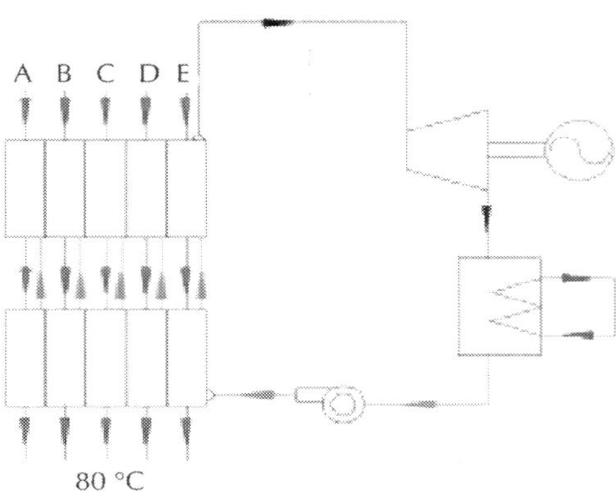

Figure 4: Schematic diagram of single ORC system.

Table 4: Simulation results of the single ORC system

Working fluid	Evaporation temperature, K	Mass flow rate, kg/s	Net power output, kW	Thermal efficiency, %
R141b	354.15	126.96	2813.02	8.63
R123	355.35	164.43	2806.48	8.61
Heptane	355.25	72.58	2795.04	8.58
R245fa	357.15	146.34	2793.94	8.58
R601	355.85	73.66	2784.73	8.55

Dual Independent ORC Subsystems with Different Working Fluids

Poor simulation results of the single ORC system indicates the necessity of two ORC subsystems, which is presented, corresponding to a high-temperature cycle and a low-temperature cycle. The evaporation temperature of the high-temperature cycle is firstly set, and by calculating the ratio of latent heat to sensible heat of the working fluid, the total heat load for this cycle is determined. Then for the low-temperature cycle, a proper evaporation temperature can be determined through interaction by the simulation program as the total heat load remaining is fixed. The heat load distribution for two ORC subsystems is shown in Fig. 5.

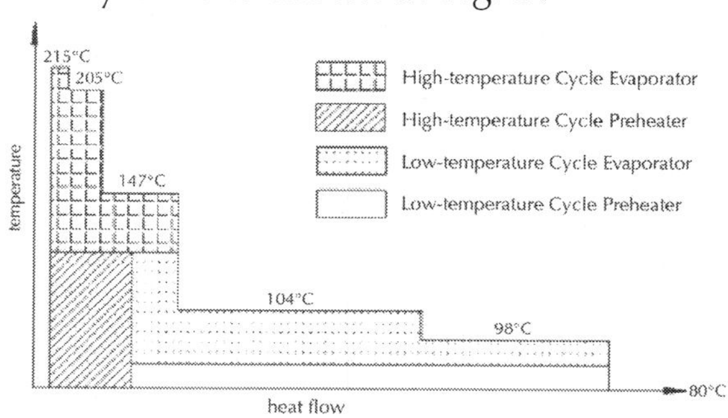

Figure 5: Heat load distribution of dual independent ORC subsystems with different working fluids.

Two different working fluids are selected for the high-temperature cycle and the low-temperature cycle, in order to achieve better performance of both subsystems. The simulation results are demonstrated in Fig. 6, which indicates that system with R245fa for the high-temperature cycle and R141b for the low-temperature cycle gives out the maximum net power output, 3353 kW, the highest among discussed schemes. Table 5shows the optimal set of thermodynamic parameters for both cycles.

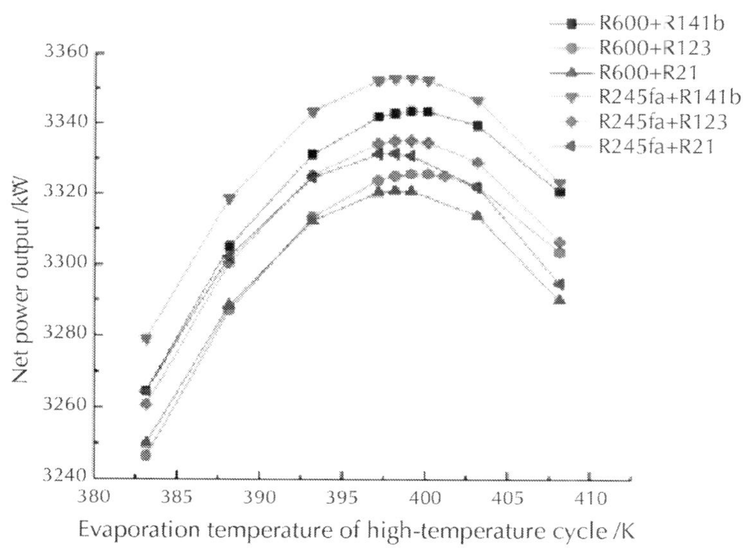

Figure 6: Simulation results of dual independent ORC subsystems with different working fluids.

Table 5: Thermodynamic parameters of dual independent subsystems with different working fluids

	Working fluid	**Evaporation temperature, K**	**Mass flow rate, kg/s**	**Net power output, kW**
High-temperature cycle	R245fa	399.15	50.7	1588.8
Low-temperature cycle	R141b	354.85	78.5	1764.5

Compared to the five independent subsystems, this scheme cuts the number of the subsystems from five to two. The entire system is relatively simplified and the efficiency is increased. However, there is still difficulty in handling two independent subsystems in operation.

Dual Integrated ORC Subsystems with One Working Fluid

Improved on the basis of the schemes above, scheme of dual integrated ORC subsystems with one working fluid is presented. The condensing temperatures of both cycles are set the same thus only one condenser and one working fluid pump are needed. The heat load distribution for the two cycles is shown in Fig. 7, and simulation results of different working fluids are shown in Fig. 8, which indicate that system with R141b as a working fluid gives out the best performance among the selected working fluids, with maximum net power output at 3325 kW, which is close to that of the scheme of dual independent subsystems and the difference is ignorable. By comparison, the scheme of dual integrated subsystems is more compact and easier to operate, thus it is regarded as the optimal scheme for this waste heat recovery case. Table 6 shows the optimal set of thermodynamic parameters for both cycles.

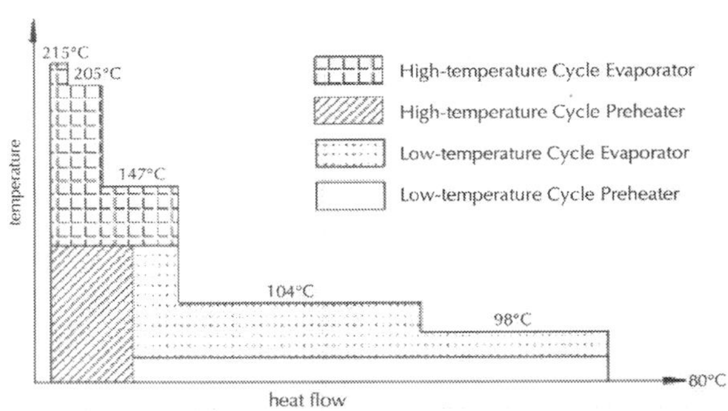

Figure 7: Heat load distribution for dual integrated ORC subsystems with one working fluid.

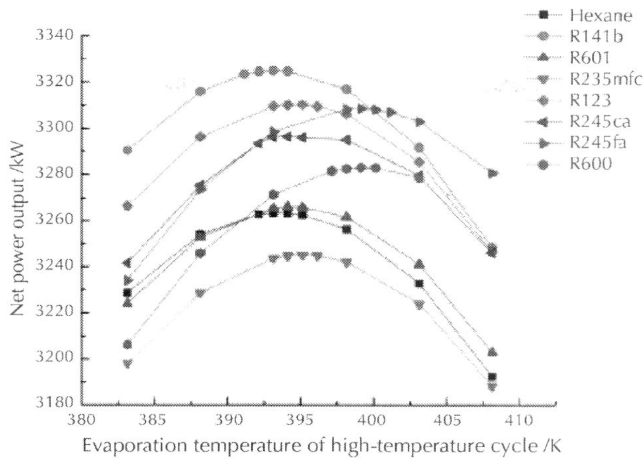

Figure 8: Simulation results of dual integrated ORC subsystems with one working fluid.

Table 6: Thermodynamic parameters of dual integrated subsystems with one working fluid

	Working fluid	Evaporation temperature, K	Mass flow rate, kg/s	Net power output, kW
High-temperature cycle	R141b	393.15	37.4	1426.3
Low-temperature cycle	R141b	353.85	86	1898.7

Schematic diagram of the recovery system including two integrated subsystems with R141b as a working fluid is shown in Fig. 9 and the flow procedure of the system is given as below: Heat sources A, B and C firstly go through the evaporator of the high-temperature cycle, and the high-grade heat load transforms the working fluid into vapor. Then A, B and a part of heat source C go through the preheater of the high-temperature cycle, making the working fluid preheated up to the evaporation temperature. Heat sources D, E and the remaining part of C go through the low-temperature subsystem in the same way, as the heat load is divided into two parts according to the ratio of latent heat to sensible heat. The distribution of the heat load of the five heat sources is listed in Table 7. It should be emphasized that all the working fluid goes into the preheater of the low-temperature cycle and then is divided into two parts, one of which goes into the evaporator of the low-temperature cycle and the other goes to the high-temperature cycle. The ratio of mass flow for the high-temperature cycle is nearly 30% of the total according to the simulation results.

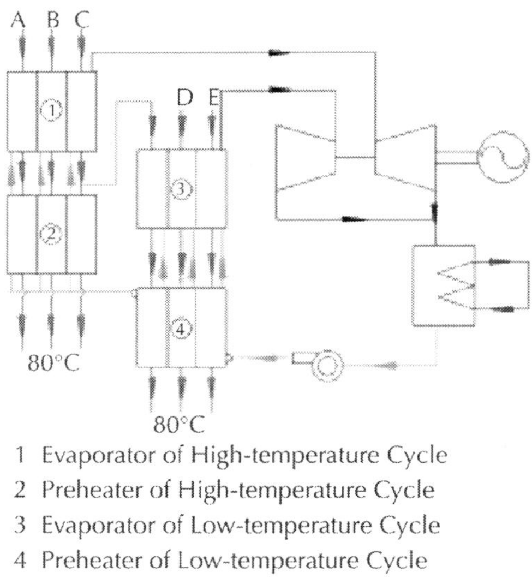

Figure 9: Schematic diagram of dual integrated subsystems with R141b as a working fluid.

Table 7: Distribution of the heat load of all the waste heat sources

Heat source	**Temperature level**	**Distribution of the heat load**			
		High-temperature cycle		**Low-temperature cycle**	
		Evaporator	**Preheater**	**Evaporator**	**Preheater**
A	215 °C	65.9%	34.1%		
B	205 °C	63.2%	36.8%		

C	147 °C	31.3%	0.8%	58.0%	9.9%
D	104 °C			72.1%	27.9%
E	98 °C			62.8%	37.2%

FURTHER ANALYSIS

The simulation results and analysis above illustrated that the system including two integrated subsystems is optimal among the schemes. In this section, further analysis is made from the view of economic factors and off-design conditions.

Economic Factors

Suppose this ORC system runs 8000 h per year, and the price of electricity is 0.08$/kWh in China, it will produce 2.671×10^7 kWh electricity, worth 2.14 million dollars. Rough estimation for the cost of ORC system would be 2500$/kW, and overall initial investment would be 8.3 million dollars. The investment recovery period would be 3.5–4 years.

Generally it is difficult to make an accurate estimation of the whole capital cost of an ORC system due to the fluctuation of the component prices. The cost of the pump and the turbine changed slightly under the condition of small power capacity level [17]. In addition, the cost of the heat exchangers such as preheater, evaporator and condenser dominates a high percentage of the overall capital cost. Thus it is reasonable that the economic analysis of the ORC system mainly involves the cost of heat exchangers, and the cost of other components can be neglected. A ratio of net power output to total heat transfer area is selected as the performance evaluation criterion for purpose of predicting the system performance from view of both thermodynamics and economics [17].

The heat transfer area is determined by

$$A = \frac{Q}{U \cdot \Delta t_m} \tag{10}$$

Where U is the overall heat transfer coefficient and t_m is the logarithmic mean temperature difference. In the computation program, the overall heat transfer coefficient is set according to the thermodynamic properties and flow state of the working fluids. The thermodynamic properties of the three organic fluids are calculated by REFROP 9.0 [12], and here the flow states are assumed the same.

Fig. 7 indicates that systems with R141b, R123 and R245fa as working fluids give out comparatively higher net power output under optimal operating condition. The ratios of net power output to total heat transfer area of systems using these three working fluids are calculated and listed in Table 8. It indicates that system with R141b as a working fluid is the most economical since its ratio of net power output to total heat transfer area, up to 0.9804, is the highest among the three working fluid candidates. There is little to none difference among the total heat transfer areas of the three systems under optimal condition. The reason is that the thermodynamic properties of the three organic fluids are similar and the flow states are assumed the same in the computation program.

Table 8: Ratios of the net power output to total heat transfer area

Working fluid	Net power output a design point, kW	Total heat transfer area, m^2	Ratio of net power output to total heat transfer area, kW/m^2
R141b	3325.03	3391.61	0.9804
R123	3310.43	3571.32	0.9269
R245fa	3308.80	3403.02	0.9723

In reality, the total heat transfer coefficient is related to many factors and is usually determined through heat transfer experiments. It remains to be a part of the future work to get a more accurate overall heat transfer coefficient for purpose of more precise analysis and further optimization.

Off-design Working Conditions

It is evident that the working condition of the reforming and extraction unit changes along with time and environment, thus the hot and cold ending conditions of the waste heat recovery system will vary. It is important to analyze the performance of the system under off-design conditions. The study is implemented for the scheme of dual integrated subsystems with R141b as a working fluid.

The mass flow rate of the waste heat sources changes in line with the product output of the reforming and extraction unit. Fig. 10 shows that the net power output of the ORC system rises as the mass flow rate of the waste heat sources increases. The net power output of the system reaches maximum when the evaporation temperature is 393.15 K in each case.

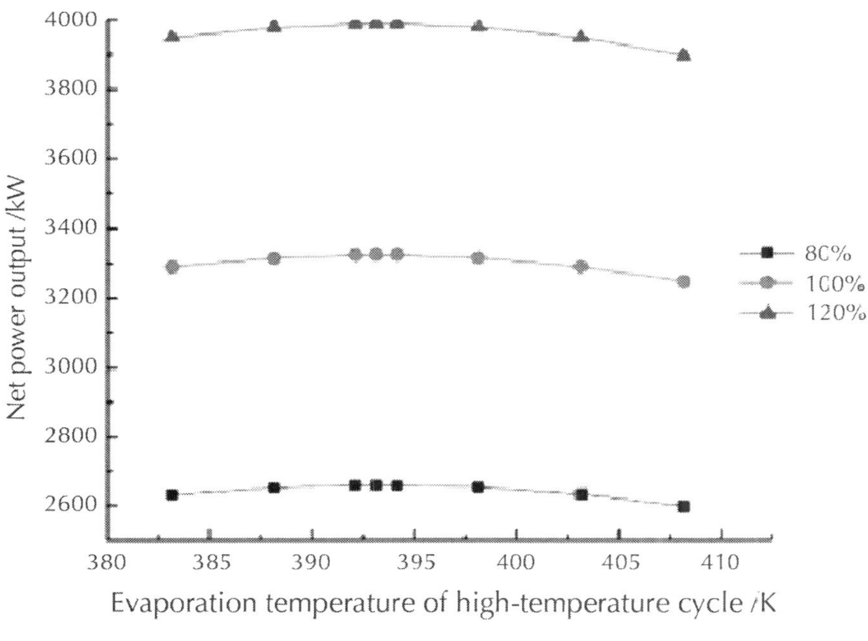

Figure 10: Net power output under different heat loads.

The temperature of the environment changes with time and it plays influence on the condensing temperature of the ORC system. Fig. 11 shows the net power output of the system under different condensing temperatures, which proves the point of view that the lower the condensing temperature, the higher the net power output. The maximum net power outputs under different condensing temperatures are shown in Fig. 12, which indicates that the influence of the condensing temperature on the high-temperature cycle is a little slighter than that on the low-temperature cycle.

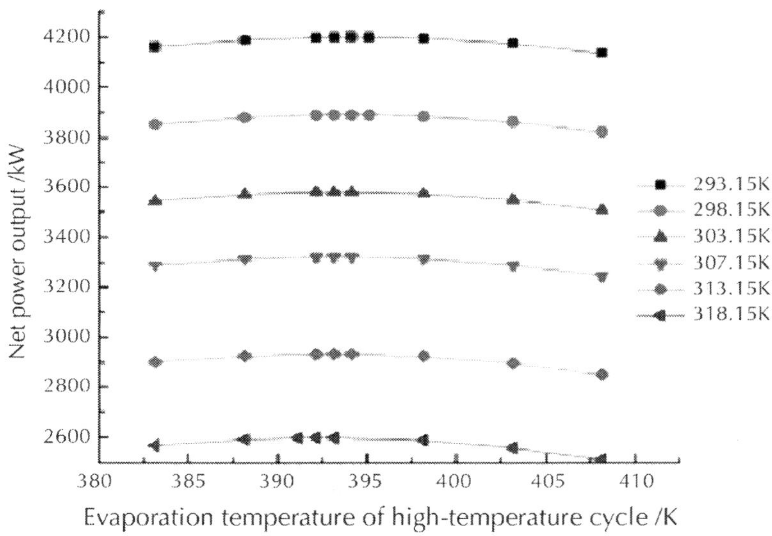

Figure 11: Net power output under different condensing temperatures.

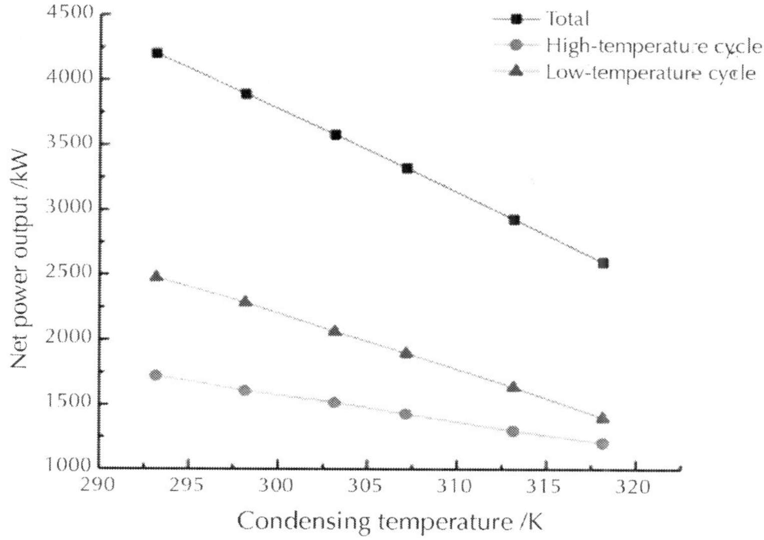

Figure 12: Maximum net power output under different condensing temperatures.

The off-design condition analysis in this paper only takes the heat load of the waste heat sources and the condensing temperature into account. In other words, all the other parameters of the system are considered the same, i.e. the turbine efficiency, the heat losses and the pressure losses and the minimum temperature differences. In the real case, the parameters mentioned above are related to the system working condition, thus it will be part of the future work to research the off-design condition more exactly.

CONCLUSIONS

This paper analyzes and optimizes a comprehensive ORC recovery system utilizing five waste heat sources distributing in different temperature levels, from a 1.2 million ton-level reforming and extraction unit in Shijiazhuang Refining & Chemical Company of China. Four schemes are presented in continuous optimization progress with the aim of simplifying the entire system and achieving both the robust thermal performance and economic benefit. The main conclusions are summarized as follows:

- The scheme of five independent subsystems is easy to design but it is difficult to handle in operation as the entire system is extremely large and complex. The maximum net power output is 3338.89 kW and the thermal efficiency is 10.25%. The scheme of single ORC system is withdrawn due to its considerably low thermal efficiency. The scheme of dual independent subsystems using R245fa and R141b as working fluids gives out the maximum net power output at 3353.3 kW. The scheme of dual integrated subsystems is more compact and the maximum net power output is 3325 kW, which is very similar with those of other schemes, thus it is recommended for this heat recovery case.
- A ratio of net power output to total heat transfer area is selected as the performance evaluation criterion from view of both thermodynamics and economics. By comparison, scheme of dual integrated subsystems R141b as a working fluid is the most recommended in this recovery case.
- Further analysis is implemented for the system of dual integrated subsystems with R141B as a working fluid, the simulation results indicate that the system performs well under off-design conditions.
- Thermal performance is the most important criterion for waste heat recovery system, and the system simplicity and technological feasibility as well as the economic factors still should be concerned for practical application. The analytical method and optimization progress proposed in this paper can be applied in similar recovery system for multi-strand waste heat sources.

REFERENCES

1. Lian HK, Li Y, Shu GYZ, Gu CW. An overview of domestic technologies for waste heat utilization. Energy Conserv Technol 2011; 29(166):123e8.
2. Bohl R. Waste heat recovery from existing simple cycle gas turbine plants e a case study. In: 18th Symposium on industrial application of gas turbines (IAGT). Alberta, Canada; 2009.
3. Saleh B, Koglbauer G, Wendland M, Fischer J. Working fluids for lowtemperature organic Rankine cycles. Energy 2007; 32(7):1210e21.
4. Roy JP, Mishra MK, Misra A. Parametric optimization and performance analysis of a waste heat recovery system using organic Rankine cycle. Energy 2010; 35(12):5049e62.
5. Sun J, Li WH. Operation optimization of an organic Rankine cycle (ORC) heat recovery power plant. Appl Therm Eng 2011; 31:2032e41.
6. Zhou N, Wang X, Chen Z, Wang Z. Experimental study on organic Rankine cycle for waste heat recovery from low-temperature flue gas. Energy 15 June 2013; 55:216e25.
7. Skarke P, Shawn M, Marcello C. waste heat recovery from internal combustion engines: feasibility study on an organic Rankine cycle with application to the Ohio state ecocar phev. In: Proceedings of the ASME 2012 Internal Combustion Engine Division Fall Technical Conference.
8. Yan Li. Design and study of low-temperature organic Rankine cycle and the turbine. Tsinghua University; 2014.
9. Hung TC, Shai TY, Wang SK. A review of organic Rankine cycles (ORCs) for the recovery of low-grade waste heat. Energy 1997; 22(7):661e7.
10. Chen H, Goswami DY, Stefanakos EK. A review of thermodynamic cycles and working fluids for the conversion of low-grade heat. Renew Sustain Energy Rev 2010; 14(9):3059e67.

11. Yamamoto T, Furuhata T, Arai N, Mori K. Design and testing of the organic Rankine cycle. Energy 2001; 26(3):239e51.
12. Liu X, Chen D, Zhang W, Qin W, Zhou W, Qiu T, et al. An assessment of the energy-saving potential in China's petroleum refining industry from a technical perspective. Energy 2013; 59:38e49.
13. Varga Z, Rabi I, Farkas C. Waste heat recovery with organic Rankine cycle in the petroleum industry. Chem Eng; 2012:29.
14. Shaomin H. Recovery of low temperature heat energy in refining using organic medium. Petrochem Des 2009; 4:022.
15. Meacher JS. Organic Rankine cycle systems for waste heat recovery in re- fineries and chemical process plants. Energy Systems Laboratory; 1981.
16. NIST Standard reference database 23. NIST thermodynamic and transport properties of refrigerants mixtures REFROP. Version 9.0; 2010.
17. Wang J, Yan Z, Wang M, Ma S, Dai Y. Thermodynamic analysis and optimization of an (organic Rankine cycle) ORC using low grade heat source. Energy 1 January 2013; 49:356e65.

Chapter 5

Application of the Exergy Concept in the Petroleum Refining and Petrochemical Industry

Ricardo Rivero

Instituto Mexicano del Petroleo, Grupo de Exergia, Eje Central Lazaro Cardenas 152, 07730 Mexico D.F., Mexico.

ABSTRACT

In this paper, the past, present and future applications of the exergy concept in applied process research, development and engineering projects in the petroleum refining and petrochemical industry, conducted at the Mexican Petroleum Institute (IMP), are described. These applications are the result of a clear recognition of the challenge represented by the reduction of energy degradation, i.e. the application of the second law of thermodynamics, through the exergy concept, which allows to simultaneously consider the three E's model: energy–economy–ecology. Applications can be grouped in two complementary directions: the first, methodological, direction consists in integrating within general simulation, analysis and optimization programs, the subroutines required to conduct exergy analyses, in order to get in a quick and precise manner, all information required to integrally optimize chemical processes through detailed exergoeconomic and exergoecologic studies. The second direction refers to research and technological development of specific systems and processes, integrally optimum, of which the justification lies on the exergy analysis, i.e. on its capacity to simultaneously consider energy, economy and ecology aspects, such as diabatic distillation, absorption heat pumps, coking–gasification-combined cycle co- and tri-generation, fuel cells, etc.

INTRODUCTION

The Exergy Concept

The increase of the energy efficiency in industry has always been one of the technical–scientific goals of managers, engineers and scientists, but the problem has now become a priority. The experience of the last three decades in the implementation of energy saving programs, has shown that it is possible to obtain up to 15% savings by first level measures, i.e. "energy conservation", avoiding wastes, like effective maintenance, increased consciousness, etc.

However in order to go further in this efforts it is necessary to apply second and third level measures, normally requiring additional investment, and for doing so, it is necessary to answer the two following questions [1]:

- Where and how is energy degraded in industry?
- How to reduce energy degradation?

As it is well known, energy cannot be created nor destroyed but only transformed (it is the first law of thermodynamics). However in all real processes its quality decreases, i.e. energy is degraded (it is the second law of thermodynamics) and the increasing relevance of the second law is the result of two significant changes produced at the international level in energy policies [2] (see Fig. 1).

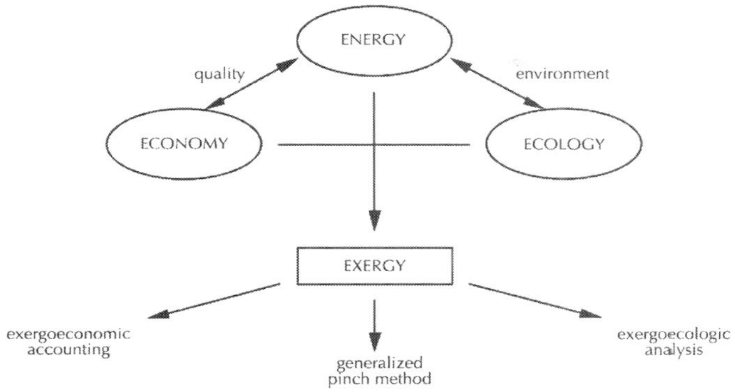

Figure 1: The energy–economy–ecology connection.

The first change corresponds to the necessity of considering not only the energy quantity when its use is to be optimized, but also its quality. This change was originated by the energy crisis of the early seventies, which brought to light the energy–economy connection and gave origin to the first energy saving programs.

The second change refers to the necessity of stopping the environment deterioration generated by the increasing fossil fuels consumption in order to assure a sustainable development. This change was originated by the recognition of the energy–ecology connection.

The exergy concept is able to deal with this three E's model because of its implications not only in terms of energy but also in terms of economy and ecology.

Exergy is calculated using the following general equation which relates enthalpy H, entropy S and temperature T:

$$\mathbf{Ex} = (\mathbf{H} - \mathbf{H}_0) - \mathbf{T}_0(\mathbf{S} - \mathbf{S}_0) \quad (1)$$

where the subscript 0 indicates the temperature, pressure, composition, etc. of the environment.

The economic implication of exergy results from the fact that exergy is a measure of the quality of energy, of its ability to perform work, thus a way of assigning costs. The first term between brackets of the right-hand side of Eq. 1, represents the total energy, the enthalpy, from which the second term, the total entropy multiplied by the temperature of the environment, is subtracted, thus obtaining the net useful energy for which a cost has to be paid. The term $T_0(S-S_0)$, represents the useless energy, the anergy, unable to perform work.

The ecologic implication of exergy results from the fact that exergy is a measure of the departure from the environment, of the difference in temperature, pressure, composition, etc. conditions with respect to the conditions of the environment, thus a way of measuring the effects, the work, that an effluent can have on the environment. The enthalpy, the entropy and the temperature in Eq. 1 are explicitly referred to the conditions of the environment.

The Mexican Petroleum Institute (IMP) works in the exergy field since the beginning of the eighties, conducting projects on the following directions:

- methodologies,
- software development,
- specific research and engineering projects and studies,
- technical and human resources.

EARLY APPLICATIONS

One of the first applications of the exergy concept at IMP was directed to the design of combustion equipment[3], [4] and [5].

During the design of the naphtha reforming units of the Tula and the Salina Cruz refineries, an exergy analysis of the reactors feed heater was conducted. This equipment, now in operation at both refineries, consists of four radiant cells, a steam generator and a steam superheater in the convection zone, a steam drum associated to that zone, an air-preheating system, and an optional water economizer.

This kind of furnace is typical in refinery and petrochemical plants, in which for many direct-fired heaters the radiation zone is used only for the process fluid and the convection zone is used for steam generation and superheating services.

The system had different design alternatives which were compared to determining the degree of efficiency in energy use and the critical points of the system. The first one was a totally radiant heater; the second included a steam generator and superheater; the third a water economizer for the steam generator; and the fourth an air-preheating system instead of the water economizer. Fig. 2 shows the fourth option.

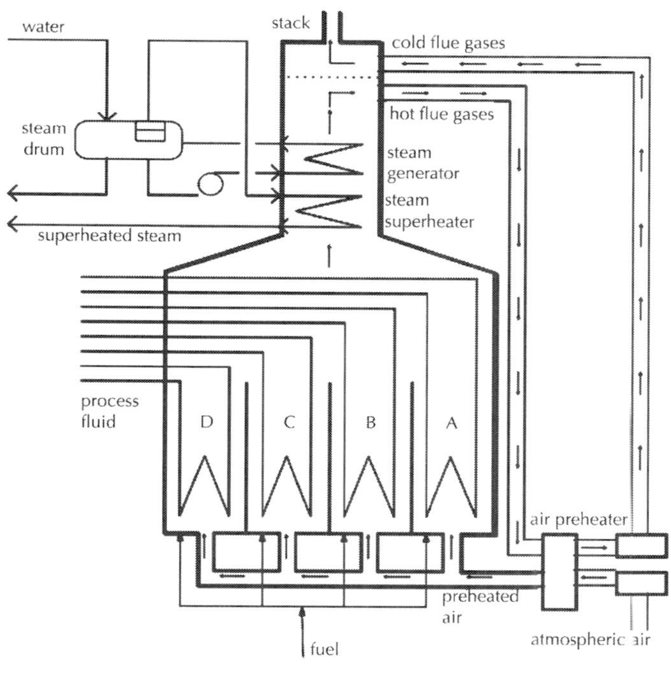

Figure 2: Reactors feed heater. Design option no. 4.

The system was divided into eight blocks for which the exergy losses were determined. From the first law (thermal efficiency) viewpoint the third and the fourth design alternatives were equally convenient, but applying the exergy balance, option 4 was better than option 3. The total exergy losses per kilogram of fuel were 18.8 MJe/kg for option 3 and 16.8 MJe/kg for option 4. This was the option chosen for the construction of this equipment. A sensitivity analysis of the preheated air temperature was conducted to determine the optimal preheater size both from the exergy and economic viewpoints. This temperature is limited by the burners construction materials, by the NO_x (oxides of nitrogen) formation, and by the SO_x (oxides of sulphur) content of the cold flue gases which may produce acid corrosion [6].

In order to conduct this exergy analysis, the existing design software named REFIMP was modified, and applied to several refining and petrochemical heaters designs [7] and [8].

From the exergy analysis of these kinds of heaters, a design criterion was obtained, that is now a common practice: steam generation with the flue gases going out of the radiation section, should be limited to the amount allowing the maximum air-preheating temperature which can be technically and economically attained.

In 1986 another relevant application of the exergy concept was made. The general process simulation code of IMP, named SIMPROC was modified to include the calculation of the physical exergy of all process streams. This simulator has been used extensively in many basic engineering projects of refinery and petrochemical plants in Mexico [9].

In 1987–1988, a project (Tula project) was conducted in order to optimize an existing 150,000 bpd (barrels per day) refinery processing a 70/30 vol% mixture of Mexican Isthmus and Maya crude oils, and consisting of the facilities listed in Table 1 which presents the main results of the exergy analysis of the refinery [10].

Table 1: Exergy analysis of an existing refinery (Tula project)

Unit	Primary energy (supplied exergy)	Useful energy (produced exergy)	Degraded energy (exergy losses)		Improvement potential (attackable losses)
			Total	Unavoidable	
Combined distillation unit	18.2	1.6	16.6	5.7	10.9
Naphtha HDS unit	2.7	0.1	2.6	1.0	1.6
Naphtha reforming unit	9.6	0.7	8.8	3.4	5.4
Intermediate distillates HDS unit	2.8	0.02	2.8	0.9	1.9
Catalytic cracking unit	18.8	2.9	15.9	6.6	9.3
Visbreaking unit	3.0	0.7	2.3	0.9	1.4
Utilities plant	44.9	12.5	32.4	13.2	19.2
Total	100.0	19.0	81.0	32.0	50.0

Table 2 shows the reduction in exergy losses obtained after the application of the improvement proposals. The increase in exergy losses for the visbreaking unit is due to the fact that the modifications proposed allowed a 100% increase in the capacity of the unit with only a 30% increase in exergy losses.

Table 2: Reduction in energy degradation of an existing refinery (Tula project)

Unit	Exergy losses (%)	
	Original	After optimization
Combined distillation unit	20.4	17.2
Naphtha HDS unit	3.2	2.7
Naphtha reforming unit	10.9	7.9
Intermediate distillates HDS unit	3.4	2.8
Catalytic cracking unit	19.5	12.7

Visbreaking unit	2.9	3.7
Utilities plant	39.8	36.7
Total	100.0	84.3

The main proposals generated in the project were:

- implementation of air-preheating systems in direct-fired heaters in the combined distillation unit, naphtha and intermediate distillates hydrodesulfurization (HDS) units, and visbreaking unit,
- heat exchangers network retrofitting by the Pinch method in the combined distillation unit, naphtha HDS unit and visbreaking unit,
- heat transfer area increase in feed-effluent heat exchangers in the naphtha reforming unit (Packinox type) and in the intermediate distillates hydrodesulfurization unit,
- relocation of the recycle hydrogen compressor in the naphtha reforming unit,
- addition of a reaction box in the visbreaking unit,
- reinstallation of the CO (carbon monoxide) combustion process and of the water economizer in the CO boiler in the fluidized catalytic cracking (FCC) unit,
- implementation of water preheaters for the boilers and a thermocompressor in the utilities unit.

Table 3 shows the evaluation of the proposals.

Table 3: Evaluation of proposals (Tula project)

Unit	**Reduction in degraded energy (recovered exergy/improvement potential) (%)**	**Proposals investment (thousands of usd)**	**Payback time (months)**	**Cost effectiveness (net present value 10 years/ investment)**
Combined distillation unit	35.9	2822.3	17	7.37

Naphtha HDS unit	24.7	1101.2	69	1.84
Naphtha reforming unit	52.7	1204.8	5	23.98
Intermediate distillates HDS unit	56.4	834.0	53	2.04
Catalytic cracking unit	59.1	7822.0	3	47 4
Visbreaking unit	–	1000.0	3	33 27
Utilities plant	13.1	660.6	4	32 19
Total	26.0	15245.0	5	24 48

Most of the proposals were implemented in the existing facilities at Tula refinery and some of them at other Mexican refineries in operation, and in the basic engineering of new plants.

In the exergy analyses conducted in the Tula project the exergy-enhanced IMP's process simulator SIMPROC was used.

Some relevant methodological aspects of that project were the exergy performance parameters used for the comparison of the different equipment composing each refinery unit in order to determine the critical points and to generate the improvement proposals for those critical points.

A distinction was made between irreversible exergy losses (Irr) and effluent exergy losses (Efl) [10] and [11]in order to explicitly use the ecological implications of the exergy concept. Also, the effectiveness ( ) instead of the efficiency ( ) was used, since the efficiency is only a different way of expressing the irreversible exergy losses whereas the effectiveness is the rational way of knowing how well the block or equipment is performing its function, i.e. the desired effect. The efficiency is defined as the ratio of the total exergy input to the total exergy output, whereas the effectiveness is defined as the ratio of the net exergy produced to the net exergy supplied.

Perhaps the most relevant aspect was the introduction of the improvement potential (Pot), which combines the irreversible exergy losses, the effluent exergy losses and the effectiveness in order to have a measure of how much and how easily the system could be theoretically improved for optimization purposes:

$$\mathbf{Pot} = \mathbf{Irr}(1 - \varepsilon) + \mathbf{Efl} \tag{2}$$

For the utilities plant of the Tula refinery, an exergoeconomic analysis was also conducted in order to determine the real production costs of electricity, and high-, medium- and low-pressure steam, supplied to the process units [10], [12] and [13]. Fig. 3 shows a comparison of the steam and electricity costs obtained by the exergoeconomic accounting technique with the values resulting from the same equations system using enthalpy values instead of exergy values. For this analysis both the physical and the chemical exergies of all streams were used. Only operation costs were considered. As shown in Fig. 3 the first law analysis would overestimate the steam costs (54% more for the low-pressure steam, 18.1% for the medium-pressure steam, and 2.7% for the high-pressure steam), and it would underestimate the cost of electricity (36% less than its real cost).

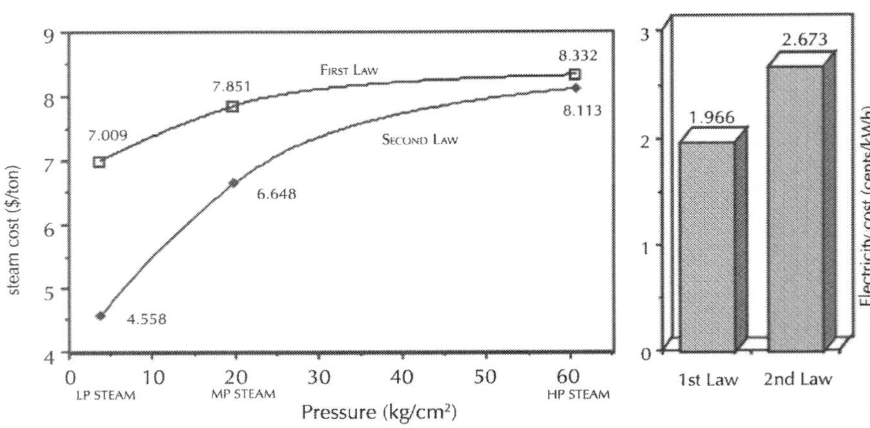

Figure 3: Electricity and steam costs based on the first and second laws. Tula project.

For the naphtha reforming unit a detailed exergoeconomic analysis using physical and chemical exergies was also conducted [14] and [15]. Fig. 4 shows the main process and utilities streams with their corresponding flow rates, monetary flows and costs per mass unit. In this analysis only transformation (raw materials and utilities) costs were considered; investment and capital costs were not involved.

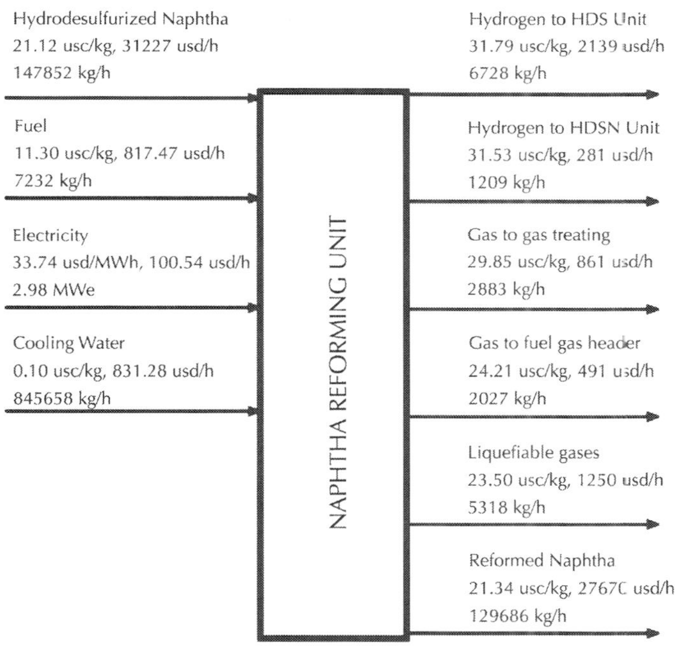

Figure 4: Main monetary flows of the naphtha reforming unit. Tula project.

In this analysis the costs of the exergy losses both irreversible and effluent were determined and an exergoeconomic improvement potential was proposed, replacing exergy values by monetary values in Eq. (2).

As shown in Fig. 4 from a single raw material, hydrodesulfurized naphtha, the naphtha reforming unit produces a series of products at different costs. The exergoeconomic accounting allows a proper allocation of production costs to each one of them.

In all analyses conducted in the Tula project all exergy performance parameters were calculated for each piece of equipment, i.e. all heat exchangers, pumps, reactors, distillation columns, compressors, turbines, etc. were individually evaluated, and some very large equipment, like combustion heaters and the main crude oil atmospheric distillation tower were subdivided into a series of blocks. Parametric analyses of the effect of the number of trays, the amount of stripping steam, the operating pressure and the use of reboilers instead of stripping steam in the three side strippers to the main tower, were conducted for the atmospheric distillation system [16] and [17]. Another important applications of the exergy concept to the optimization of refining and petrochemical existing facilities were conducted for the lubricants train of the Salamanca refinery [18], to the ammonia plant of the Cosoleacaque petrochemical complex [19], and to the methanol plant of the Texmelucan petrochemical complex [20].

RECENT APPLICATIONS

During 1989–1993, a co-operative research project on the application of exergy analysis to diabatic (non-adiabatic) distillation and absorption heat pumps was conducted at the Laboratory of Chemical Engineering Sciences (LSGC) in Nancy, France, directed by Prof. Pierre Le Goff, and co-financed by Atochem, of the French petroleum and petrochemical group Elf-Aquitaine, by the French Agency of Energy and Environment (ADEME), by the IMP and by the National Science and Technology Council of Mexico (CONACYT) [21].

Methodologies

On the methodological side, the main applications of the exergy concept of that period (1989–1993) were directed to distillation processes and absorption heat pump systems. The generalized pinch method, i.e. the enthalpy–Carnot temperature diagram was applied to the analysis of both classical (adiabatic) and diabatic distillation [21], [22] and [23].

Diabatic distillation is a process in which heat is transferred inside the column reducing energy degradation (exergy losses) in a considerable (up to 50%) amount. Fig. 5 shows an example of adiabatic and diabatic distillation columns for the rectification of a water–ethanol (EtOH) mixture, and Fig. 6 shows the corresponding enthalpy–Carnot temperature diagram. The total exergy losses of the columns, represented by the surface between the exergy source and sink curves in the diagram, are less that 60% for the diabatic column compared to the adiabatic column.

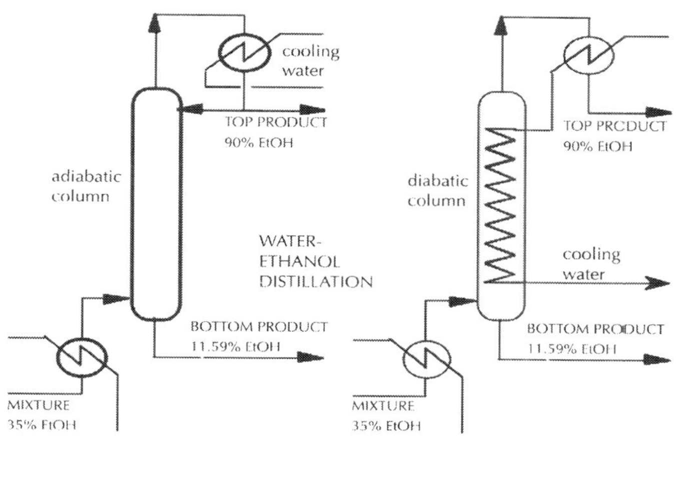

Figure 5: Adiabatic and diabatic columns.

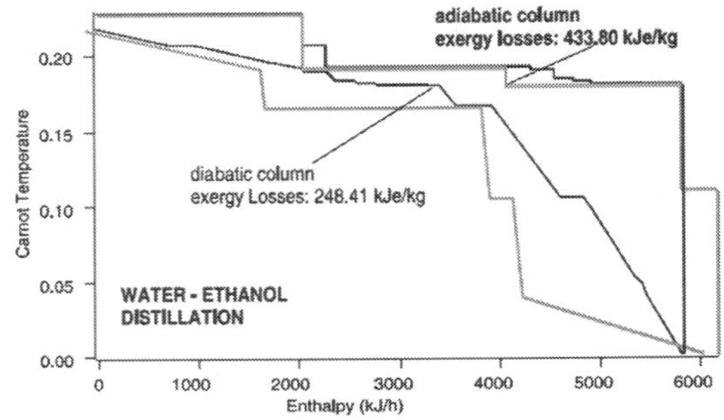

Figure 6: Enthalpy–Carnot temperature diagram.

Another application of the exergy concept to the analysis of heat and mass transfer processes was the development of tri-dimensional exergy diagrams, as the one shown in Fig. 7, which was applied to the exergy analysis of absorption heat pump systems [21] and [24].

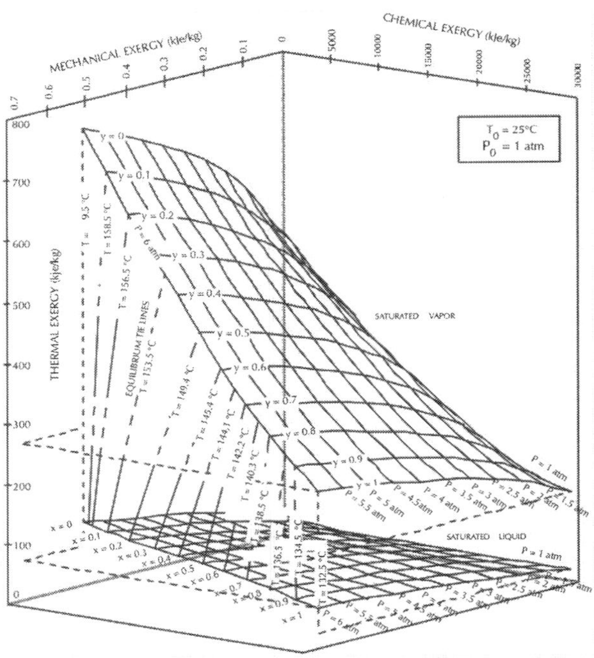

Figure 7: Tri-dimensional exergy diagram.

In order to go further in the exploitation of the ecological implications of the exergy concept, a performance parameter called the exergoecologic improvement potential was proposed [21] and [25]; this parameter could be used for calculating an exergy tax for industrial installations.

Also a life cycle analysis based on the total (irreversible and effluent) cumulative exergy losses, extending the approach of Szargut et al. [26] was proposed [21].

Finally, an overall classification of heat pumps based on the exergy multi-pole concept was proposed [27].

Advanced Energy Systems

An experimental installation of diabatic distillation was constructed and operated in the frame of the research project [21], [28] and [29]. This experimental study demonstrated the technical feasibility of diabatic distillation and its advantages from the exergy losses viewpoint. The internal structure of the diabatic distillation trays is shown in Fig. 8. They have a very simple geometry which could be used to diabatize conventional existing distillation columns.

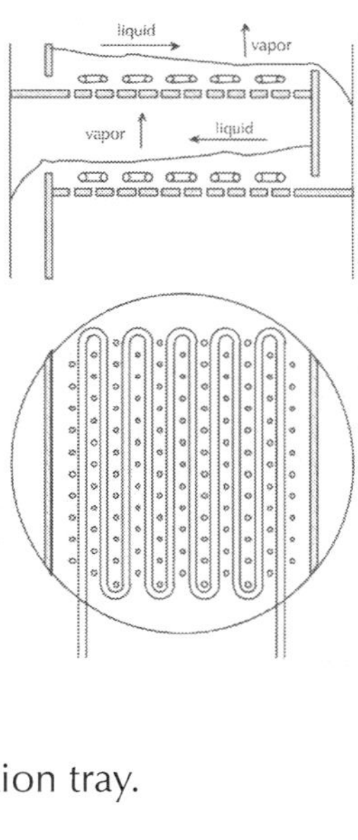

Figure 8: Diabatic distillation tray.

Diabatic distillation was also studied by computer simulation in order to determine the optimal distribution of heat to be transferred inside the column [21], [30], [31], [32] and [33].

Finally, the application of diabatic columns as components of absorption heat pumps as the system shown inFig. 9 was proposed [21] and [34]. The system has a thermal coefficient of performance (Cop) of 0.45 and an effectiveness of 0.38.

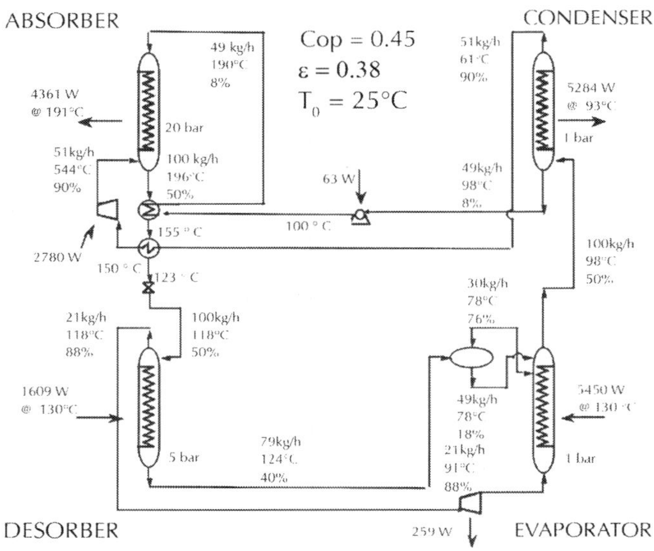

Figure 9: Diabatic distillation heat pump.

PRESENT AND FUTURE APPLICATIONS

Fig. 10 shows the present applied process research directions of the IMP exergy group. The application of the exergy concept in the refining and petrochemical industries is being conducted in two general directions.

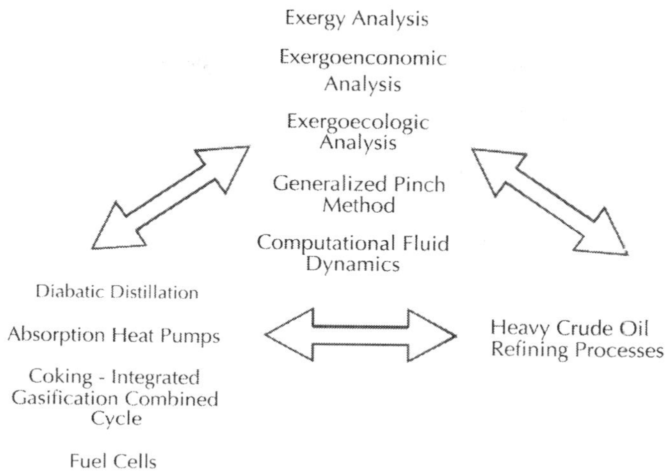

Figure 10: Applied process research directions of the IMP exergy group.

First, the technological development of methodologies by incorporating exergy analysis to the existing simulation, analysis and optimization tools in order to conduct exergoeconomic and exergoecologic studies directed to locate inefficiencies and establish their economic consequences, to determine the real production costs of the processes, and to evaluate their environmental impact.

Second, the technological development of advanced energy systems like: diabatic distillation processes in refinery and petrochemical units, heat pumps and heat transformers systems in refinery units, coking and integrated gasification combined cycle units for heavy crude oils, fuel cell systems in refinery and petrochemical units.

Exergoeconomic Analysis

A 270,000 bpd refining scheme for the processing of a 50/50 vol% mixture of Mexican Maya and Isthmus crude oils has been studied [35]. The fundamental objective of the project was the integral (energy–economy–ecology) optimization of a whole heavy crude oil refining scheme by conducting a detailed exergy analysis of the situation of the advanced refining processes which are being implemented by Pemex, determining the monetary flows in all points of the processes and adequately allocate the overall production costs to the different refinery products, gasoline, kerosene, gas oils, etc., and sulfur, steam and electricity, as well as generated effluents, considering not only their quantity but mainly their quality and ecologic implications.

The project was divided in three parts: simulation, exergy analysis and exergoeconomic analysis which are being conducted for the refinery units listed in Table 4.

Table 4: Refinery units considered in the exergoeconomic analysis project

Unit
Combined distillation unit
TAME unit
Visbreaking unit
Alkylation unit
Naphtha HDS unit
Hydrocarbons treatment and fract. unit
Naphtha reforming unit
Combined cycle co-generation unit
Isomerization unit
LPG oximer treatment unit
Jet fuel HDS unit
Sulfur recovery unit
Jet fuel oximer treatment unit
Hydrogen unit
Kerosene HDS unit
Delayed coking unit
Diesel deep HDS unit
Gas oil HDS unit
Fluidized catalytic cracking unit
Air separation unit
Propane–propylene unit

Coke gasification unit
MTBE unit

For this project, a series of subroutines have been coupled to the AspenPlus simulation code for calculating physical, chemical and total exergy values of all the process and utilities stream, and the exergy performance parameters of all equipment components, including irreversible, effluent and total exergy losses, effectiveness, improvement potential, etc., [35], [36] and [37]. Once all simulation, exergy analysis and exergoeconomic analysis modules are completed for all units, the whole refinery scheme is being integrated in the IMP Cray computer. In parallel the determination of the exergoecologic improvement potential of each unit is being determined, obtaining the pollution factors of all material effluents of the refinery units [25]. At present, the six large refining complexes of Mexico (Cadereyta, Madero, Salamanca, Tula, Minatitlán and Salina Cruz) are being reconfigured, to increase the processing of Maya crude oil from 32 to 57 vol% by year 2002, to achieve a processing capacity of 771,000 bpd of Maya crude oil of a total of 1.345 millions of bpd (1.565 installed capacity) [38]. The simulation, exergy analysis and exergoeconomic analysis infrastructure will be completed before that date and it will be a useful tool for the energy–economy–ecology optimization of the industrial facilities.

Diabatic Distillation

The application of diabatic distillation to the many distillation systems of refineries and petrochemical plants is being investigated. The exergy losses of these systems represent from 30% to 60% of the total exergy losses in many process units. At present three distillation systems have been studied: the desbutanizer and the desisohexanizer towers of the naphtha HDS unit, and the naphtha stabilization tower of the naphtha reforming unit; these units are located at the Cadereyta refinery, but similar facilities exist in all other refining complexes.

Fig. 11 shows the diabatic distillation system of the naphtha stabilization tower. It produces 65% of the exergy losses, uses 16% of the cooling water, and has 6% of the reboiler heat duty of the adiabatic distillation system[39] and [40].

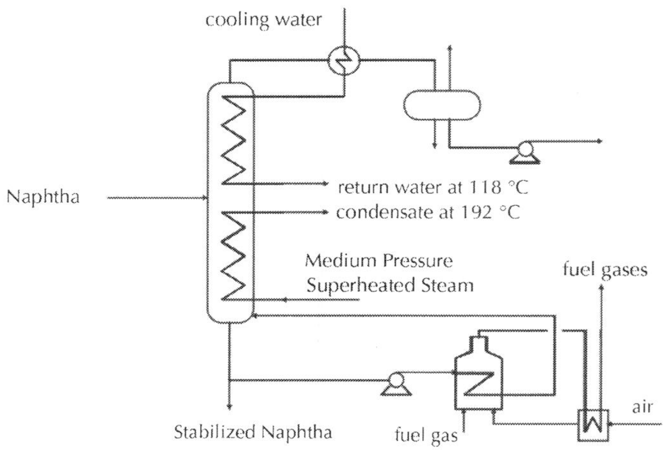

Figure 11: Diabatic distillation system for naphtha stabilization.

The research work in this field will be oriented to the diabatization of an existing tower probably inside the Salamanca refinery, and in the evaluation on a pilot plant scale of different geometries of vapor–liquid contacting devices with incorporated heat transfer.

Reactive Distillation

The results of the exergy analysis of the methyl ter butyl ether (MTBE) unit of the Salina Cruz refinery indicate that the main exergy losses (about 55%) of the plant occur in the reactive distillation system, particularly in the distillation column itself and in its associated air-cooled condenser [35] and [41].

A detailed exergy analysis of the reactive distillation system has been conducted in order to improve the operation.

The effect on the exergy losses and on the MTBE yield, has been evaluated for changes in the location of the methanol feed to the column, variations in the excess methanol flow rate, and modifications in the reboiler heat duty. Fig. 12 shows a simplified diagram of the system.

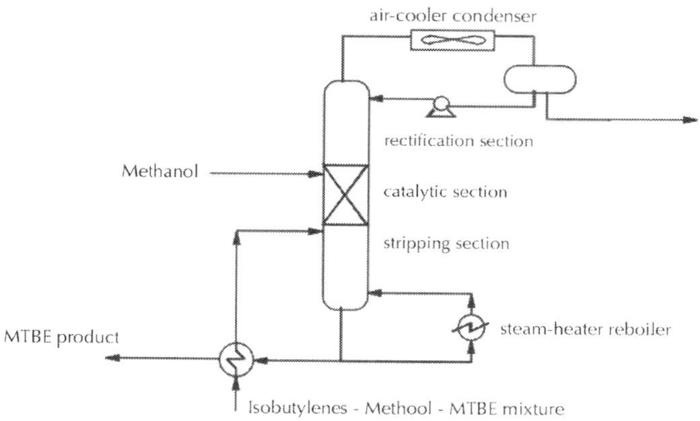

Figure 12: Reactive distillation system for MTBE production.

As an example of the kind of results obtained from the exergy analysis of the reactive distillation system it can be said that when the methanol feed is located in the rectification section, a minimum of exergy losses is obtained but also a near-to-the-minimum MTBE production. However when the methanol feed is located near to the bottom of the catalytic section, a near-to-the-minimum exergy losses and a near-to-the-maximum MTBE production are obtained; this is an adequate feed location from both points of view [41].

Absorption Heat Pumps

Currently five applications of absorption heat pumps to refining and petrochemical plants are being investigated: (1) crude oil desalting water purification, (2) condensable gases recovery, (3) assisted distillation, (4) ammonia refrigeration, and (5) trigeneration systems.

The design of a pilot plant for an absorption heat pump assisted distillation system is expected to start next year.

The potential for upgrading waste heat in sector 1 of the Tula refinery has been determined. Also a preliminary evaluation of a water–lithium bromide heat transformer coupled with the desisopentane distillation system of the new isomerization unit under construction at Cadereyta refinery, has been completed, showing a potential energy saving of 45% of the reboiler heat duty [35], [42] and [43]. Fig. 13shows the assisted distillation system.

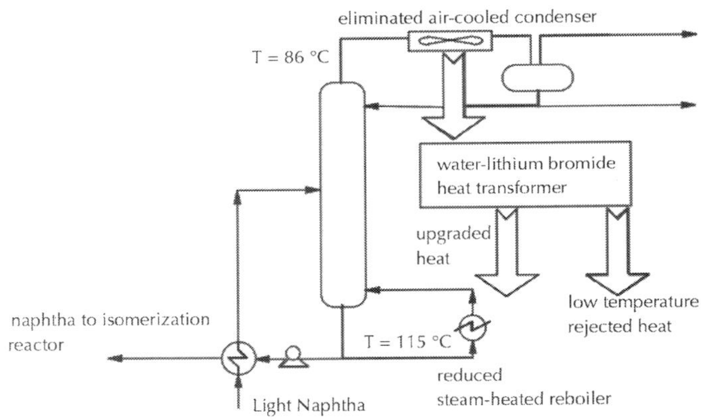

Figure 13: Heat transformer assisted distillation system in the isomerization unit.

This distillation system belongs to the hydrocarbon feed preparation section of the isomerization unit and represents about 54% of the total exergy losses of the unit, 51% from which correspond to the reboiler and the condenser of the tower, hence the potential for heat pumping.

The application of an ammonia–lithium nitrate absorption heat pump for an ammonia plant at the Cosoleacaque and Camargo petrochemical complex is being investigated and a pilot plant of the refrigeration system is expected to be constructed in the site next year [44].

Coking–integrated Gasification Combined Cycles

In Cadereyta, Madero and Minatitlán refineries, three new reduced crude oil delayed coking units are under construction, in order to increase the production of gas oil for FCC for gasoline formulation [38]. These units will produce important amounts of petroleum coke, which could be gasified and then used in a co-generation combined cycle [35] and [45]. Fig. 14 shows the proposal which is being investigated. An air separation unit is needed to produce nitrogen for coke treatment and oxygen for the gasification process. The hydrogen and carbon monoxide gas obtained is then treated for sulfur removal and cooled; then it is used in a gas turbine for electricity production. Flue gases from the gas turbine produce high-pressure steam in a heat recovery steam generator which is then used in a steam cycle co-generation system to produce more electricity and the heat required by the refinery process units.

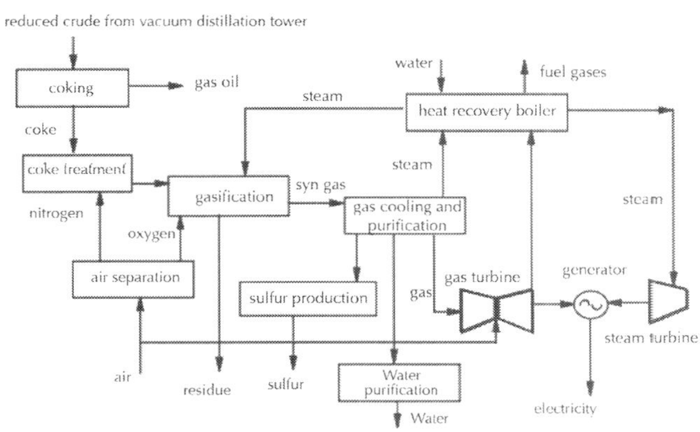

Figure 14: Reduced crude oil coking–integrated gasification combined cycle.

The exergoeconomic analysis of this process will provide detailed information on the production costs of electricity, sulfur, and steam at different pressure levels, of the refinery schemes.

Fuel Cells

Fuel cells technology is developing very rapidly and cost effective large scale systems are expected to be available in a few years. They use hydrogen as the primary energy vector for producing electricity and heat by an oxidation without combustion process [46].

In crude oil refineries and petrochemical plants, there are hydrogen producing units and hydrogen consuming units. Producing units are mainly naphtha reforming, benzene–toluene–xylene (BTX) reforming and FCC units, and consuming units are mainly naphtha, gas oil, jet fuel, kerosene and diesel HDS units. For many years the hydrogen balance in refineries was positive, i.e. there was an hydrogen excess which was unfortunately incorporated to the refinery fuel gas header and burned in combustion equipment. In the last years hydrotreating units forced the inclusion of methane steam reforming units in refinery schemes.

As in the coking-IGCC (integrated gasification combined cycle), the exergoeconomic accounting technique will provide the required information for the implementation of fuel cell electricity (and heat) generation units in refinery schemes like the one proposed in Fig. 15.

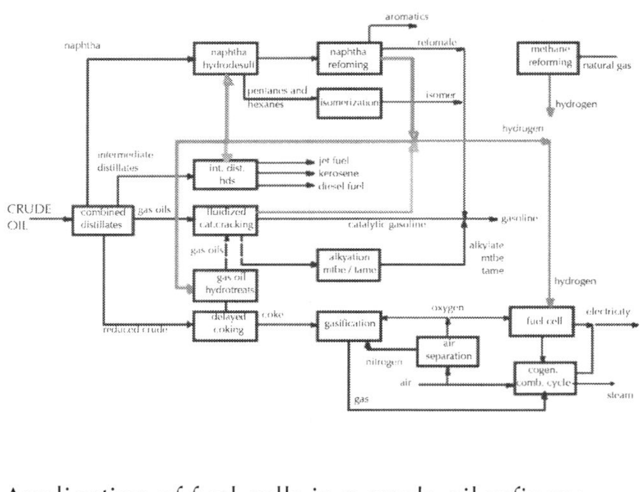

Figure 15: Application of fuel cells in a crude oil refinery.

The refinery scheme shown in Fig. 15 includes a gasification process for a combined cycle. Depending on the gasification technology, the syngas could also provide hydrogen for the fuel cell or for the hydrotreating process units of the refinery. The naphtha reforming unit is used more for hydrogen production than for reformate because of the benzene content of this gasoline component. Finally a gas methane reforming unit for hydrogen production could also be considered. All process scheme options are to be analyzed by the exergoeconomic accounting technique.

CONCLUSIONS

In energy-intensive sectors of industry, like the refining and petrochemical industry, the relevance of the exergy concept lies in its ability to consider in an integral way the three E's model: energy–economy–ecology. Many applications of the exergy concept have been made and exergy based technics will continue to be used for the improvement of energy use in industry, which will help in reducing energy degradation in a technically feasible, costly effective and ecologically sustainable way. The exergy concept provides more, and more detailed, information on the performance of industrial processes for their improvement and optimization.

REFERENCES

1. Le Goff P. Energetique Industrielle, tome 1 ISBN 2-85206-053-1. Paris: Technique et Documentation, 1979.
2. Rivero R. Situacion Actual y Tendencias Tecnologicas en Conservaci on y Ahorro de Energ a. En: Memorias de la II Reunion de Trabajo para la Determinaci on del Programa de Investigaci on y Desarrollo Tecnologico a Mediano y Largo Plazos. Mexico: Pemex-Refinacion, 1995. p. 13
3. Rivero R, Garcıa-Price R. El Analisis de Exergia Aplicado al Diseno de Calentadores a Fuego Directo con ~ Sistemas

Integrados de Precalentamiento de Aire, En: XXV Convencion Nacional del IMIQ, Inst. Mex. de Ing. Quim., San Luis Potosı, 1985. p. 44.

4. Anaya A, Rivero R. Application of the concept of exergy to the optimum utilization of energy in industrial processes and equipment. Proceedings of World Congress III of Chemical Engineering, vol. I. Tokyo: Asian-Pacific Conf. of Chem. Eng., Interamerican Conf. of Chem. Eng., European Fed. of Chem. Eng., 1986. p. 407–10.
5. Rivero R. Exergy analysis of a naphtha reforming heater. IX International Congress on Chem. Engineering, Chem. Equipment Design and Automation, CHISA '87, 353rd Event of The European Fed. of Chem. Eng., Praha, Czecoslovakia, 1987. p. 10.
6. Rivero R. Sistemas de Precalentamiento de Aire en Calentadores a Fuego Directo. Revista del Instituto Mexicano del Petroleo ISSN: 0538-1428 1986;17(4):45–65.
7. Rivero R. REFIMP3: Programa de Diseno de Hornos de Carga a Reformadores de Nafta. M ~ exico: Instituto Mexicano del Petroleo, 1984.
8. Rivero R, Garcıa-Price R. EXERGY: Programa de Evaluacion de Exergia para CAFD con SIPA. M exico: Instituto Mexicano del Petroleo, 1985.
9. IMP, SIMPROC: Simulador General de Procesos. Mexico: Inst. Mex. Petroleo, 1987.
10. Rivero R, coordinador tecnico. Optimizacion de las Instalaciones Existentes en la Refinerıa de Tula, Hidalgo, Utilizando el Metodo de Exergia. Reporte Final del Proyecto E-2277, Tomos I a IX. Mexico: Grupo de Exergia, Inst. Mex. del Petroleo, 1988. p. 713
11. Rivero R, Gonzalez G, Garcıa V, Pulido R, Escarcega C. Exergy analysis of a crude oil atmospheric distillation unit. In: Cai RX, Moran MJ, editors. Thermodynamic Analysis and Improvement of Energy Systems ISBN: 0-08- 037516-2.

Beijing: International Academy Publishers, Pergamon Press; 1989, p. 506–11.

12. Rivero R, Hernandez R, Ortega A, Barrios H. Analisis Exergoeconomico de un Sistema de Cogeneraci on. En: IV Pacific Area Chem. Eng. Congress, PACHEC '88. Acapulco: Interamerican Conf. Chem. Eng., Asian-Pacific Conf. Chem. Eng. 1988. p. 10.
13. Rivero R, Hernandez R. Exergoeconomic analysis of a refinery's utilities plant. In: Sciubba E, Moran MJ, editors. 2nd Law Analysis of Energy Systems: Towards the 21st Century ISBN: 88-86662-00-9. Roma: Centro Internazionale di Ricerca e Calcolo Universitario Scientifico, ASME; 1995, p. 325–34.
14. Rivero R, Escarcega C, Nunez A, Terreros P, Led ~ on F. An alisis Exergoeconomico de una Unidad Reformadora de Naftas. En: IV Pacific Area Chemical Engineering Congress, PACHEC '88. Acapulco, Mexico: The Interamerican Conf. Chem. Eng., The Asian-Pacific Confederation of Chemical Engineering, 1988.
15. Rivero R. Analisis Exergoeconomico de Procesos. Revista del IMIQ ISSN: 0188-7319 1995;36(7–8):19–24.
16. GarcıaV, Gonzalez G, Pulido R, Rivero R. Analisis Exergico de una Torre de Destilacion Atmosf erica de Crudo. Revista del IMIQ 1988. Revista Especial XXVIII Convencion Nacional. p. 23–32.
17. Rivero R, Anaya A. Exergy analysis of a distillation tower for crude oil fractionation. In: Tsatsaronis G, Bajura RG, Kenney WF, Reistad GM, editors. Computer-Aided Energy-Systems Analysis ISBN: 0-7918-0582-4, AES-vol. 21. New York: American Society of Mechanical Engineers; 1990, p. 55–62.
18. Grupo de Exergia. Estudio Tecnico Economico entre el Uso de Venas de Vapor y Trazadores El ectricos en Lineas de Proceso, para Plantas de Tula, Hgo. y Salamanca, Gto., Proyecto E-2277. Mexico: Grupo de Exergia, Inst. Mex. del Petroleo, 1988. p. 16.

19. Grupo de Exergia. Optimizacion de Plantas Petroqu ımicas Utilizando el Metodo de Exergia, Proyecto E-2292: Planta de Amoniaco, Cosoleacaque, Ver. Mexico: Grupo de Exergia, Instituto Mexicano del Petroleo, 1989. p. 111.
20. Grupo de Exergia. Optimizacion de Plantas Petroqu ımicas Utilizando el Metodo de Exergia, Proyecto E-2292: Planta de Metanol, San Martın Texmelucan, Puebla. Mexico: Grupo de Exergia, Instituto Mexicano del Petroleo, 1989. p. 116.
21. Rivero R. L'Analyse d'Exergie. Applicationa la Distillation Diabatique et aux Pompes a Chaleur a Absorption. These de Doctorat, Institut National Polytechnique de Lorraine, Nancy, 1993. p. 680.
22. Rivero R, de Oliveira Jr S, Le Goff P. Temperature de Carnot h, Diagramme h/Q, Methode du Pincement. Entropie ISSN: 0013-9084 1990;26(157–158):13–20.
23. Le Goff P, Rivero R, de Oliveira jr S, Cachot T. Application of the enthalpy–Carnot factor diagram to the exergy analysis of distillation processes. In: Tsatsaronis G, Gaggioli RA, El-Sayed YM, Drost MK, editors. Fundamentals of Thermodynamics and Exergy Analysis ISBN: 0-7918-0580-8, AES-vol. 19. New York: American Society of Mechanical Engineers; 1990, p. 21–8. 12
24. Rivero R. Tridimensional exergy diagram. In: Szargut J, Kolenda Z, Tsatsaronis G, Ziebik A, editors. Energy Systems and Ecology, vol. 1. Cracow: Polish Ministry of National Education, American Society of Mechanical Engineers; 1993, p. 305–12.
25. Rivero R. The exergoecologic improvement potential of industrial processes. In: Cai RX, Moran MJ, Zhang SZ, Xiao YH, editors. Thermodynamic Analysis and Improvement of Energy Systems ISBN: 7-5062-3264-Z/TK6. Beijing: World Publishing Corporation; 1997, p. 299–304.
26. Szargut J, Morris DR, Steward FR. Exergy Analysis of Thermal, Chemical and Metallurgical Processes ISBN 0-89116-574-6. New York: Hemisphere Publishing Corporation; 1988.

27. Le Goff P, Rivero R. A thermal inventory of sorption heat pumps based on the exergy-multipole concept. In: Gog us YA, Ozt urk A, Tsatsaronis G, editors. Efficiency, Costs, Optimization, Simulation and Environmental Impact of Energy Systems ISBN: 975-7475-07-6, vol. 1. Istanbul: Middle East Technical University, American Society of Mechanical Engineers; 1995, p. 190–4.

28. Rivero R, Cachot T, Ramadane A, Le Goff P. Diabatic or quasi-reversible distillation, exergy analysis, industrial applications. In: Kouremenos DA, Tsatsaronis G, Rakopoulos CD, editors. Analysis of Thermal and Energy Systems. Athens: Greg. Foundas Technical and Scientific Editions, ASME; 1991, p. 129–40.

29. Rivero R, Cachot T, Le Goff P. Exergy analysis of adiabatic and diabatic distillation columns: an experimental study. In: Pilavachi PA, editor. Energy Efficiency in Process Technology ISBN: 1-85861-019-2. London: Elsevier Applied Science Publishers Ltd; 1993, p. 1254–67.

30. Rivero R. Exergy simulation of a distillation tower: adiabatic rectification column. In: Tsatsaronis G, editor. Engineering Systems Design and Analysis Design of Energy Systems, PD-vol 64-3 ISBN: 0-7918-1280-4, vol. 3. New York: ASME; 1994, p. 1–12.

31. Rivero R. Exergy simulation of a distillation tower: adiabatic stripping column. In: Krane R, editor. Thermodynamics and The Design, Analysis, and Improvement of Energy Systems 1994 ISBN: 0-7918-1455-6, AES-vol. 33. New York: ASME; 1994, p. 135–45.

32. Rivero R. Exergy simulation of a distillation tower: diabatic rectification column. In: Sciubba E, Moran MJ, editors. 2nd Law Analysis of Energy Systems: Towards the 21st Century ISBN: 88-86662-00-9. Roma: Centro Internazionale di Ricerca e Calcolo Universitario Scientifico, ASME; 1995, p. 465–76.

33. Rivero R. Exergy simulation of a distillation tower: diabatic stripping column. In: Gog us YA, Ozt urk A, Tsatsaronis G, editors. Efficiency, Costs, Optimization, Simulation and Environmental Impact of Energy Systems ISBN: 975-7475-07-6, vol. 1. Istanbul: Middle East Technical University, ASME; 1995, p. 163–7.
34. Rivero R, Le Goff P. Pompes a Chaleur a distillation diabatique. In: Auracher H, Feidt M, Tsatsaronis G, editors. Thermodynamics, Heat and Mass Transfer of Refrigeration Machines and Heat Pumps. Nancy: Institut National Polytechnique de Lorraine; 1998, p. 227–33.
35. Rivero R. investigador en jefe. Optimizacion Ex ergica de un Esquema de Refinacion, Proyectos E0A-7442, P.00257, F.00257, Mexico: Grupo de Exergia, Instituto Mexicano del Petroleo. 1996–2000.
36. Rivero R, Daroux M, Benali T. Calculation of thermal, mechanical and chemical exergy of material streams using a process simulation program. In: Gog us YA, Ozt urk A, Tsatsaronis G, editors. Efficiency, Costs, Optimization, Simulation and Environmental Impact of Energy Systems ISBN: 975-7475-07-6, vol. 1. Istanbul: Middle East Technical University, American Society of Mechanical Engineers; 1995, p. 168–72.
37. Rivero R, Rendon C, Monroy L. Calculation of the exergy of crude oil mixtures. In: Bejan A, Feidt M, Moran MJ, Tsatsaronis G, editors. Efficiency, Costs, Optimization, Simulation and Environmental Aspects of Energy Systems and Processes ISBN: 2-905-267-29-1. Nancy: Institut National Polytechnique de Lorraine; 1998, p. 211–8.
38. Willars JM. Reconfiguracion del Sistema Nacional de Refinaci on. Revista del IMIQ ISSN 0188-7319 1998;39 (11–12):31–36.
39. Rivero R. investigador en jefe. Investigacion, y Desarrollo Tecnol ogico de Procesos de Destilaci on Diab atica. Proyecto D.00018. Programa de Investigacion en Medio

Ambiente y Seguridad (PIMAS), M exico: Grupo de Exergia, Instituto Mexicano del Petroleo. proyecto en desarrollo.

40. Rivero R, Monroy L. Analisis Exergico de una Torre de Destilacion Diab atica para la Estabilizacion de Naftas. En: XXXVIII Convencion Nacional del IMIQ. Oaxaca, M exico: Instituto Mexicano de Ingenieros Quımicos, 1998.

41. Rivero R, Garcia M. Exergy Analysis of a Reactive Distillation MTBE Unit. International Journal of Applied Thermodynamics ISSN 1301-9724 2001;4(2):85–92.

42. Rivero R, Urquiza J. Perspectivas de Aplicacion de Bombas de Calor en Procesos de Refinaci on. en: XXXVIII Convencion Nacional del IMIQ. M exico: Inst. Mex. Ing. Quim. Oax. 1998.

43. Rivero R, Best R. coordinadores. Bombas de Calor por Absorcion en los Procesos de Destilaci on del Petr oleo. Informe Final del Proyecto FIES-96-33-II, Mexico: Grupo de Exergia, Instituto Mexicano del Petroleo/Centro de Investigacion en Energ ıa, Universidad Nacional Autonoma de M exico, 1999.

44. Rivero R, investigador en jefe. Estudio de Optimizacion Energ etica de la Planta de Amoniaco de Petroquımica Camargo, S.A. Mexico: Grupo de Exergia, Instituto Mexicano del Petroleo. proyecto en desarrollo.

45. Rivero R. Opciones Tecnologicas para el Coque Producido por una Unidad de Coquizaci on de Residuo de Vac ıo. Boletın Monitec 1997;4(5):6–7.

46. Rivero R. Las Celdas de Combustible y sus Perspectivas de Aplicacion en Procesos de Refinaci on de PetroleoRevistadelIMIQ ISSN:0188-7319 1997;38(11–12):18–26.

Chapter 6

An Assessment of the Energy-saving Potential in China's Petroleum Refining Industry from a Technical Perspective

Xiaoyu Liu[a, c], Dingjiang Chen[a], Wenjun Zhang[a], Weizhong Qin[a], Wenji Zhou[a, d], Tong Qiu[a], and Bing Zhu[a, b]

[a]Department of Chemical Engineering, Tsinghua University, Beijing 100084, PR China

[b]International Institute for Applied Systems Analysis, Schlossplatz 1, A-2361 Laxenburg, Austria

[c]Department of Civil and Environmental Engineering, University of Maryland, Collage Park, 20742 MD, United States

[d]Petroleum Company Ltd., China National Aviation Fuel Group, Beijing 100088, PR China

ABSTRACT

As a major contributor to productivity and employment in China, the petroleum refining industry consumes approximately 15% of industrial fuel oil and 10% of industrial coal. Given this energy-intensive characteristic, cost-effective investments for energy-efficient technologies may be a useful strategy to improve the competitiveness of China's refining industry. More importantly, this approach may alleviate the environmental problems China faces. This paper addresses the challenges posed by a highly complex refining system, incomplete industrial information, and the absence of a widely accepted evaluation method. This study models and analyzes the energy-savings potential for refining and conversion processes in the context of technological change. The results indicate that upgrading process heaters have been a priority during recent years, but heat recovery and advanced process control systems will gradually begin to dominate the technological marketplace in the long term. Current technology policies will result in approximately 2.7×10^8 GJ of energy savings by 2020, keeping the average energy consumption of refineries within 57 kg oil equivalent (kgoe)/t-feed. If a cap-and-trade scheme is introduced in the future, a further reduction of up to 10% can be achieved. Various specific barriers that impede the realization of potential goals are also addressed in this study.

INTRODUCTION

The petroleum refining industry in China is the second largest such industry in the world and provides products to many other sectors, including the transportation and chemicals industries. As of January 2010, China's petroleum refining industry operated 54 refineries and yielded profits in excess of 1800 billion RMB (Renminbi, Chinese Yuan) in 2009 [1] and [2]. China's petroleum refineries must expand to accommodate growing urbanization and the concomitant rise in transportation demand. Sixty percent of a refinery's operating costs are typically spent on energy. Enhanced energy efficiency helps refineries reduce costs as well as environmental pollution and greenhouse gas emissions.

The petroleum refining industry consumes approximately 15% of industrial fuel oil and 10% of industrial coal [3]. Energy use in refineries fluctuates due to changes in the types of crude that are processed, the product mix and complexity of the refinery, and the sulfur content in the final products [4]. In 2010, the energy consumption of refineries was equivalent to approximately 30 million tons of oil, which is approximately 1.5 times more than the energy consumption ten years ago and twice as much as fifteen years ago [5] and [6]. Energy consumption will continue to increase because of trends toward more complex processing systems with better conversion capacities. China's government has implemented policies to facilitate energy management practices in petroleum refineries [7], [8] and [9]. Over the past decade, the CNPC (China National Petroleum Company) and SINOPEC (China Petroleum and Chemical Corporation) have decreased their unit energy consumption by 28% and 34%, respectively [2] and [5]. Nevertheless, their energy consumption is still 13% higher than that of U.S. refineries (53 kgoe/t-feed), indicating significant room for improvement in energy consumption [10] and [11]. It is necessary to clarify the potential for improvement with regard to the effectiveness and cost of available technology to implement policies and attract relevant investment.

In addition, the use of energy-efficient technology will also meet the challenges posed by modern refineries that consist of more complex and integrated systems. Historically, Chinese refineries were only capable of distillation and lacked reforming or converting capacities in accordance with domestic low-sulfur crude. The adoption of energy-efficient technologies only addressed problems with distillation heaters and the control of the reflux ratio. In the early 1990s, China became a net importer of high-sulfur crude oil and began taking steps to modernize refining systems that not only distill hydrocarbon compounds but also convert and blend them into a wider array of products [4] and [12]. Today, technological opportunities are no longer limited to distillation but have expanded to oil upgrading, conversion, treating, and blending as well as systems for the optimal control of an integrated refinery [13].

The United States and Brazil have implemented research programs to use comparative benchmark data to quantify technological opportunities in energy efficiency improvements of petroleum refineries [4], [14], [15], [16] and [17]. Their results showed the potential improvements that could be realized if refineries adopted all available commercial energy-efficient technologies over the next decade. The potential benefits were most likely overestimated when investment returns, market competition, and policy restriction were not considered. Johansson et al. investigated European petroleum refineries that could save 0.17 GJ (GJ) per year by using short-term mitigation options, such as fuel substitution and energy efficiency measures [18]. The estimated energy-saving potential for the individual abatement strategies depended on strong assumptions made in the analysis. The results were subject to considerable uncertainty with respect to the choice of technology and scope of implementation. Park et al. created a system dynamics model that incorporated a basic energy/raw material flow and process flow to evaluate the energy saving potential of the petroleum refining industry in Korea [19]. This model focused on the change in production capacity but ignored the impact of cost-effectiveness on technology adoption. Existing bottom-up energy demand models (e.g., MARKAL, DNE21+, AIM/end-use and PRIMES) explicitly consider the stock turnover based on a single objective cost criterion and are well-suited to identify promising technology options in complex systems at the regional level. However, these models have, to the best of our knowledge, not been applied to analyses of potential energy savings in the Chinese petroleum refining industry.

Based on the current body of literature, we attempted to develop an optimization model that accounts for the factors that affect technology adoption. To do so, we combined a bottom-up approach with technology diffusion theory to predict how technology will be implemented in the future and to assess potential energy savings. A bottom-up approach was used to establish the relationship between technology adoption, energy savings, costs, production plans, and policy constraints. Previous research on technology selection in other industrial sectors (e.g., lime, pulp and paper) has proven the applicability of this approach [20], [21], [22] and [23]. In contrast to previous research, this study contributes an industrial bottom-up energy efficiency model based on the following aspects: 1) detailed real-life technological parameters from Chinese refineries and industrial data of the entire sector's composition as of 2010; 2) explicit application of energy-efficient technologies in Chinese refineries. These applications are convenient for the users to consider the future role of energy efficiency improvements in the overall Chinese energy strategy; 3) a detailed introduction of the carbon policy instrumented under China's cap-and-trade scheme. This introduction reflects the decision-making criteria and quantifies the impacts of the emission cap and price on pathway selection in the Chinese petroleum refining industry.

This paper is organized as follows: Section 2 presents the overall structure of the optimization model and the acquisition of input parameters. A cap-and-trade policy scenario is introduced into this model, which helps to quantify the impacts of the carbon cap and price on pathway selection in the Chinese petroleum refining industry. Section 3 predicts the technology diffusion trends for six principle refining processes, analyzes their potential energy savings, and simulates the uncertainties using the Monte Carlo technique. It also presents how this potential changes under future cap-and-trade schemes and carbon emission reductions that result from improved energy efficiency. Section 4 concludes with implications of this model for the design of economically feasible and environmentally friendly technology regulations.

METHODOLOGY

An Overview of the Model

To build a model that combines a bottom-up approach with technology diffusion theory, we begin by selecting principle refining processes and the practical technologies that are applicable to the six processes as the crucial parts of the entire system. The model is designed to optimize the cost-benefit ratio of the refining process and prioritize technologies that will help refineries meet energy conservation objectives while accounting for capacity and technology adoption constraints. The model is presented in Fig. 1. We subsequently describe the classification and acquisition of the model's input parameters.

- Bottom-up module: Based on the bottom-up modeling approach, this module performs a structural analysis of the refining process, selects key energy-efficient technologies, and matches energy-efficient technologies with refining processes based on the operation flowchart.
- Optimization module: For each refining process, dynamic technological structures can be modeled as optimal solutions by maximizing the ratio of annual energy savings to technological costs that meet national energy conservation objectives. The expected result is determined by two categories of variables. The first category of variables is introduced capacity ($x_{n,t}$), which drives increases in the annual installation capacity. The other variable category is the technology diffusion coefficient ($\alpha_{n,i,t}$, $\beta_{n,i,t}$), which describes the diffusion of technological information through two channels: 1) autonomous acquisition, where potential adopters take the initiative to use new technologies and 2) influence by previous technology adopters. The probability of autonomous acquisition, , is referred to as the external effect coefficient. The probability of influence by previous adopters, , is referred to as the internal effect coefficient.

- Input module: Two types of parameters are required to support the model›s operation. Sectoral parameters are used to predict changes in the installation capacity over the next decade (constraint 1). Energy conservation requirements (constraint 2) and financial measures of energy savings (objective) are also quantified. Technological economic parameters apply to the measurement of energy savings, investment costs, and market share for each specific technology that are necessary for the model›s objective and all the constraints.
- Output: The direct output of the model simulates technology diffusion trends for six principle refining processes to calculate potential energy savings. A Monte Carlo simulation was used to study the output distribution of optimal solutions and evaluate changes in potential energy savings for the petroleum refining industry.

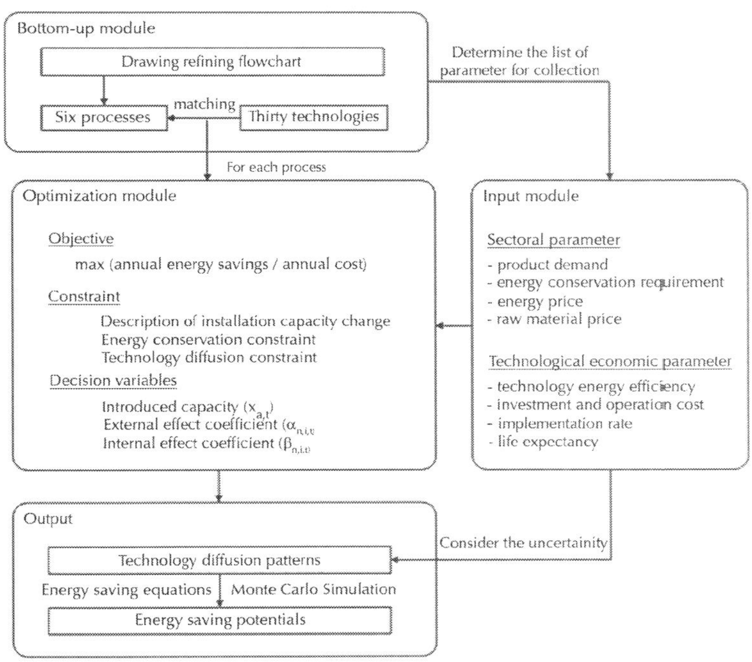

Figure 1: The structure of potential analysis model.

Boundary Definition and Opportunity Identification

The petroleum refining industry includes establishments that refine crude petroleum into refined petroleum products through multiple, distinct processes [16]. This paper examines the potential savings for six principle energy-consuming processes: 1) atmospheric and vacuum crude distillation units (ADU and VDU), 2) FCC (fluid catalytic crackers), 3) hydrocrackers, 4) hydrotreaters, 5) catalytic reformers, and 6) cokers (Fig. 2). These processes account for approximately 70% of the energy consumed by the refining industry and offer significant opportunities to increase energy efficiency [24].

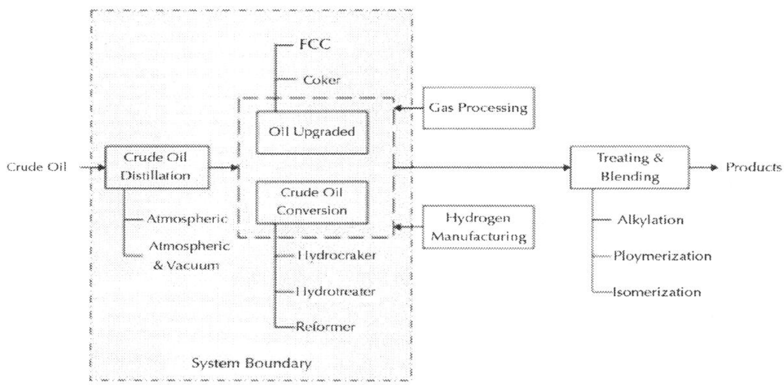

Figure 2: Simplified flowchart of refining processes and boundary definition.

Forty to sixty percent of the opportunities for energy efficiency improvement could be realized by implementing energy-efficient technologies [28]. These technologies can be broken down into six major categories: 1) energy recovery, 2) heat exchangers, 3) process heaters, 4) cross-cutting equipment, 5) process changes and design, and 6) process control systems. Table 1 lists specific practices for each of the major energy-efficient technology categories that are applicable to the six principle energy-consuming processes.

Because of the extreme complexity, it is impossible to include all of the energy-saving practices in this analysis. Although new technologies are continuously being developed, we only focus on the technologies that are commercially available in China.

Table 1: Matrix of energy efficiency opportunities in petroleum refineries

	ADU & VDU	**FCC**	**Hydrocracker**	**Hydrotreater**	**Reformer**	**Coker**
Energy recovery	Blowdown steam recovery Water reuse Heat pipe technique	Flare gas recovery Boiler economizer Reduce surface losses Blowdown steam recovery	Hot water tracing Hydraulic turbine	Waste heat recovery from quadruple furnace		Low level heat utilization Water reuse
Heat exchangers	Vertical plate and frame exchanger Medium water heater	Fouling mitigation Vertical plate and frame exchanger			Fouling mitigation	Site pinch analysis Vertical plate and frame exchanger
Process heaters	Draft control Air preheating	Shock wave soot blower Corrugated plate air pre-heater		New burners		Draft control Corrugated plate air pre-heater •Air preheating
Cross-cutting equipment	Variable speed drives Compressor throttling controls	Variable speed drives	Compressor throttling controls			

Process changes and design	Higher efficiency desalting	Stripper optimization Higher efficiency feed nozzle	Heat high-pressure separating process	Improved catalyst		Second condensation
Process control systems	Distributed control system		Advanced process control (APC)	Distributed control system	Advanced process control (APC) Distributed control system	

Model Design

The potential energy savings for each refining process can be modeled as the solution to a non-linear function. The objective function maximizes the ratio of annual energy savings to technological costs in the context of a dynamic technological structure. This calculation must 1) not exceed the given refining capacity for each process and technology, 2) meet the government's requirements for energy efficiency improvements, and 3) adopt new technologies according to the logistic equation under technology diffusion theory. The objective function and constraints are described in detail below.

Objective Function

The objective of this optimization model was to select the most cost-effective technological combinations for each refining process for the next decade. The objective of this optimization model for a certain process,n, given a target year, t, is described in Equation (1). The decision variables are in brackets under the "max" operator and include the newly introduced capacity ($x_{n,t}$), external effect coefficient ($\alpha_{n,i,t}$), and internal effect coefficient ($\beta_{n,i,t}$), which will be illustrated in the constraints. For six refining processes, the total number of variables xn,t, $\alpha_{n,i,t}$, $\beta_{n,i,t}$ are 36, 222 and 222, respectively.

$$\max_{\substack{\{x_{n,i},\alpha_{n,i},\beta_{n,i}\} \\ n\in\{1,\dots,N\};\ i\in\{1,\dots,I\}}} \quad Z_{n,t} = \frac{\Delta ES_{n,t}}{\Delta C_{n,t}} = \frac{ES_{n,t} - ES_{n,t-1}}{C_{n,t} - C_{n,t-1}} \tag{1}$$

$ES_{n,t}$ represents the energy savings for the refining process, n, in year, t, expressed in MJ. $C_{n,t}$ represents the technological costs of n in year t expressed in RMB. The two variables can be calculated using the following equations:

$$ES_{n,t} = \sum_{i=1}^{I} ES_{n,i,t} = \sum_{i=1}^{I} ue_{n,i,t} \cdot q_{n,t} \cdot r_{n,i,t} \tag{2}$$

$$C_{n,t} = \sum_{i=1}^{I} C_{n,i,t} = \sum_{i=1}^{I} \left(f_{n,i} + oc_{n,i}\right) \cdot q_{n,t} \cdot r_{n,i,t} \tag{3}$$

$$f_{n,i} = I_{n,i} \cdot \left(1 - sr_{n,i}\right) \cdot \frac{(1+\gamma)^{T_n} \cdot \gamma}{(1+\gamma)^{T_n} - 1}, \quad oc_{n,i} = \sum_{e=1}^{E} eu_{n,i,e} \cdot p_e \tag{4}$$

For a technology i in process n in a given year t, $ue_{n,i,,t}$ is the unit savings from the overall energy requirement expressed in MJ/t, $q_{n,t}$ is the capacity of the refining process n expressed in tons, $r_{n,i,t}$ is the implementation rate, $f_{n,i}$ is the unit annual investment cost expressed in RMB/t, $oc_{n,i}$ is the unit operation cost expressed in RMB/t, $I_{n,i}$ is the unit initial investment cost expressed in RMB/t, $sr_{n,i}$ is the subsidy rate that is equal to 0 without relevant policies, γ is the discount rate that is specified as 0.1 when market barriers are not considered (a higher discount rate can be used to partly account for market uncertainty and an exogenous limit for the diffusion speed of new technologies), T_n is the installation lifetime for the refining process n expressed in years, $eu_{n,i,e}$ is the consumption of energy e when using technology i expressed in t/t (kWh/t for electricity), and p_e is the price of energy e expressed in RMB/t (RMB/kWh for electricity).

Description of Installation Capacity Change

Equation (5) describes the substitution process for the refining facilities. When the installations reach the end of their lifetime (T_n), they retire and new capacities ($x_{n,t}$) should be introduced to meet the yearly growing demand [23]. Assuming that the initial capacity is $q_{n,t0}$ and that xn,k is introduced for each year k, the retired capacity for a certain year t should be given by $\left(q_{n,t_o} + \sum_{k=1}^{t-1} x_{n,k}\right) / T_n$. Given the capacity in the previous year, $q_{n,t-1}$, in the target year t, the total capacity, $q_{n,t}$, should be defined as follows:

$$q_{n,t} = q_{n,t-1} - \left(q_{n,t_0} + \sum_{k=1}^{t-1} x_{n,k}\right) / T_n + x_{n,t} \tag{5}$$

Energy Conservation Objective Constraints

China's Ministry of Environmental Protection has instituted cleaner production standards to regulate energy consumption in the petroleum refining industry [7]. Energy conservation objectives that must be met in the next five years are assigned to primary refining processes. The following constraints ensure that the annual energy savings growth rate, η_n (with a value ranging from 1% to 3%), meets or exceeds the cleaner production standard requirements for each refining process:

$$\Delta ES_{n,t}/ES_{n,t} \geq \eta_n \tag{6}$$

Technology Diffusion Constraints

Technology diffusion theory considers the adoption of new technologies driven by information dissemination, which often complies with a logistic function [25]. As discussed in Section 2.1, two channels for the dissemination of technological information are considered in this study. For a certain technology, i, used for process n in target year t, the probability of implementing this technology by potential users is $\alpha_{n,i,t} + \beta_{n,i,t} \cdot n_{n,i,t}$ if there are already $n_{n,,t}$ adopters for a total of $N_{n,i,t}$.[1] After a given time, Δt, the increase in the number of technology adopters is equal to $\Delta n_{n,i,t}+\Delta t = (\alpha_{n,,t} + \beta_{n,i,t} \cdot n_{n,i,t})(N_{n,i,t} - n_{n,i,t})\ \Delta t$. Thus, for the target year t, the implementation rate of technology, i, in process n is given by Equation (7), in which δ is a constant coefficient that can be obtained using the value of $r_{n,i,0}$, $\alpha_{n,i,0}$, $\beta_{n,i,0}$ and $N_{n,i,0}$ (the value of these parameters/variables at the year of 2010):

$$r_{n,i,t} = \frac{n_{n,i,t}}{N_{n,i,t}} = \frac{-\alpha_{n,i,t-1}/N_{n,i,t-1} + \exp\delta(\alpha_{n,i,t-1} + \beta_{n,i,t-1}N_{n,i,t-1})}{\beta_{n,i,t-1} + \exp\delta(\alpha_{n,i,t-1} + \beta_{n,i,t-1}N_{n,i,t-1})} \tag{7}$$

It is rare for a single technology to exclusively dominate the market, so the value for $r_{n,i,t}$ is limited by the upper value of $r_{n,i}^{max}$. $r_{n,i}^{max}$ is often set to 0.9, but is reduced for particular technologies to avoid any possible overlap or conflicts. The probability that potential adopters obtain information about a certain technology is affected by its relative cost-effectiveness and is also limited by communication technologies and cooperative behavior. Therefore, the value ranges for $\alpha_{n,i,t}$ and $\beta_{n,i,t} \cdot n_{n,i,t}$ are defined according to their maximum (0.8) and the relative magnitude of unit energy savings for the technology specified, as shown in the following equations:

$$0 \le \alpha_{n,i,t} \le 0.8 \times \left(\frac{\Delta ES_{n,i,t}}{\Delta C_{n,i,t}}\right) \Big/ A_n \tag{8}$$

$$0 \le \beta_{n,i,t} \le (0.8/n_{n,i,t}) \times \left(\frac{\Delta ES_{n,i,t}}{\Delta C_{n,i,t}}\right) \Big/ A_n \tag{9}$$

$$A_n = \max_{i \in \{1,\ldots,I\}} \frac{\Delta ES_{n,i,t}}{\Delta C_{n,i,t}} \tag{10}$$

Scenario for Co_2 Emission Reduction

China has not yet set a national cap on its greenhouse gas emissions, citing the need to grow its economy. Instead, it has set future limits on carbon intensity–a first-time goal in the ongoing 12th-five-year plan to reduce CO_2 intensity (CO_2 emissions per unit of GDP (gross domestic product)) by 17% from the 2010 levels by 2015 [34]. The petroleum refining industry is one of the most energy-intensive sectors with the largest CO_2 emissions due to its excessive energy demand for distillation and cracking processes and increasingly large volume. It is important for petroleum refining industry to reduce its greenhouse gas emissions in the near future. Therefore, this scenario is designed to assess the impacts of the CO_2reduction policy instrumented in this industry in China.

Abatement Target and Permit Allocation

The CO_2 abatement targets for the entire industrial sector are set to "returning the CO_2 emission to year 2005 level by 2050" based on the research of Dai et al. [35]. Their research suggested that this target could be achieved by stabilizing CO_2 emissions Chinese industry and provided the emission reduction targets through energy efficiency improvement for 2020, 2030 and 2050. The change in targets from 2010 to 2020 is assumed to be linear. This assumption is advantageous because targets advance gradually. The annual emission permits are defined as the emission in the previous year minus the emission reduction target (Table 2), and the annual permits for sub-sectors are allocated based on the annual emission targets.

Table 2: Emission permits at selected time points over the planning horizon

Year	2010	2012	2014	2016	2018	2020
GT-CO_2	6.50	6.86	7.25	7.68	8.15	8.66

With regard to permit allocation methods, Lee et al. [44] analyzed the impact of permit allocation methods on carbon reduction and accumulated GDP loss at the industrial level. Their result indicated that the "grandfathering rule", namely allocating the permits by sectoral emission proportions in a base year, was a more feasible approach to allocate permits to each industry sector. Since the current five-year plan for emission caps began in 2010, CO_2 emission permit allocation principles are formulated with Equations(11) and (12), taking 2010 as the base year:

$$RQ_t^P = EQ_t^P \cdot \frac{RQ_{2010}^A}{EQ_{2010}^A} \tag{11}$$

$$Q_{n,t}^P = RQ_t^P \cdot \frac{Q_{n,2010}^A}{RQ_{2010}^A} \tag{12}$$

where RQ_t^P and $Q_{n,t}^p$ are the emission permits for the entire refining industry and refining process, respectively, in year t . EQ_{2010}^A is the amount of CO_2 emitted by the entire industrial sector in 2010, which is 6.5 Gt according to Dai et al. [35]. RQ_{2010}^A and $Q_{n,2010}^A$ are the amounts of CO_2 emitted by the refining industry and refining process, n, in 2010, which are calculated based on the energy consumption reported in previous studies [3] and [11]. EQ_t^p represents the CO_2 emissions cap for the entire industrial sector in year t as listed in Table 3.

Table 3: Energy savings, cost and implementation rate of thirty energy-efficient technologies

No.	Technology	Unit savings (MJ/t-feed)		Unit cost (RMB/t-feed)	
		Average	Changing range	Average	Changing range
1	Fouling mitigation	26.6	[18.6, 34.6]	0.65	[0.51, 0.79]
2	Vertical plate and frame exchanger	22.7	[20.9, 24.5]	1	[0.44, 1.56]
3	Corrugated plate air pre-heater	43.7	[30.6, 56.8]	0.32	[0.19, 0.45]
4	Draft control	17.6	[15.7, 19.5]	0.4	[0.29, 0.51]
5	Air preheating	50.2	[25.1, 75.3]	1.2	[0.40, 2.00]
6	Variable speed drives	13.4	[6.7, 20.1]	0.15	[0.12, 0.18]
7	Shock wave soot blower	262.3	[183.6, 341.0]	8.1	[4.86, 11.34]
8	New burners	188.1	[126.0, 250.2]	0.45	[0.30, 0.60]
9	Hydraulic turbine	20.9	[10.2, 31.6]	29.3	[26.96, 31.64]
10	Compressor throttling controls	29.7	[25.5, 33.9]	2.85	[1.71, 3.99]
11	Hot water tracing	126.2	[103.5, 148.9]	0.05	[0.03, 0.07]
12	Higher efficiency desalting	12.1	[6.1, 18.2]	1.75	[0.81, 2.70]

13	Higher efficiency feed nozzle	39.7	[17.5, 61.9]	2.2	[0.59, 3.81]
14	Stripper optimization	26.8	[15.3, 38.3]	0.1	[0.05, 0.15]
15	Improved catalyst	62.7	[42.0, 83.4]	4.25	[0.81, 7.69]
16	Heat high-pressure separating process	11.1	[7.8, 14.4]	25	[15.00, 35.00]
17	Second condensation	33.4	[29.1, 37.7]	1.1	[1.00, 1.20]
18	Water reuse	0.2	[0.1, 0.3]	0.12	[0.11, 0.13]
19	Blowdown steam recovery	10	[9.6, 10.4]	0.03	[0.02, 0.04]
20	Medium water heater	50.2	[33.6, 66.8]	0.65	[0.40, 0.90]
21	Heat pipe technique	31.4	[25.1, 37.7]	0.75	[0.20, 1.30]
22	Reduce surface losses	418	[359.5, 476.5]	3.1	[1.86, 4.34]
23	Steam generation through waste heat recovery	40.5	[27.1, 53.9]	0.7	[0.42, 0.98]
24	Boiler economizer	188.1	[167.4, 208.8]	2.9	[2.00, 3.80]
25	Flare gas recovery	104.5	[83.6, 125.4]	5	[4.75, 5.25]
26	Waste heat recovery from quadruple furnace	50.4	[37.8, 63.0]	7.5	[4.50, 10.50]
27	Low level heat utilization	121.2	[83.6, 158.8]	1	[0.60, 1.40]
28	Site pinch analysis	106.8	[85.4, 128.2]	1.5	[0.90, 2.10]
29	Distributed control system	6.3	[4.4,8.2]	0.25	[0.17, 0.33]
30	Advanced process control	12.5	[3.9,21.1]	0.6	[0.20, 1.00]

Emission Quantification

CO_2 emissions during refinery operations are proportional to the amount of energy consumed and the emission factor of each type of energy. The CO_2 emission factor for each energy form, e, is represented bycece. For fuel oil, cec_e is 77.4 $kgCO_2e/GJ$ [40]. For natural gas, cec_e is 2.2 $kgCO_2e/Nm^3$ [45]. For electricity,cec_e is 1.0 $kgCO_2e/kWh$ [41]. Therefore, the total amount of CO_2 emitted by a refining process, n, in year t can be calculated as follows:

$$Q^A_{n,t} = \sum_{i=1}^{I} \sum_{e=1}^{E} eu_{n,i,e} \cdot cec_e \tag{13}$$

Cap-and-Trade Scheme

To investigate the impact of a combination in carbon capping and emissions trading on the refining industry, the base scenario and the cap-and-trade scenario are framed to simulate the adoption of energy-efficient technologies and the investment return on energy efficiency improvement. The base scenario is defined as the adoption of current energy conservation policy. This scenario is used as the criterion for comparing the reduction capacity and cost impact of future carbon reduction measures. For the cap-and-trade scenario, cap-and-trade policy fixes the total amount of emissions and applies a market mechanism to reduce carbon [32] and [33]. If the actual amount of CO_2 emissions is higher than the cap in a year, the cost to cover the excess CO_2 emissions for that year must be considered. Conversely, if the actual amount of CO_2 emissions is below the cap, the sector can then obtain revenue from selling the surplus amount of CO_2 reduction. The CO_2 price (CP_t, price per unit CO_2 emissions) is assumed to continually rise on an annual basis beyond 2016. The CP_t for 2016 is assumed to be 50 RMB/t-CO_2, with an estimated annual increase of 5% between 2016 and 2020 [44]. Therefore, the paid cost of CO_2 emissions (or revenue gained) (EC_t)for the refining industry in year t can be calculated with Equation (14). The total cost of the refining process, n, in year t in Equation (3) is the sum of the emissions cost ($EC_{n,t}$) and the technological cost ($C_{n,t}$).

$$EC_{n,t} = CP_t \cdot \left(Q^A_{n,t} - Q^P_{n,t}\right) \tag{14}$$

Data Collection and Uncertainty Analysis

The model involves nearly 500 parameters that fall into two categories. The first category includes sectoral parameters, which are the major external factors that affect future technology adoption, such as product demand, energy requirements, and the price of energy and materials. The major sources for these parameters are the relevant sectoral standards, policy guidelines, and consulting reports [7], [9] and [26]. The second category includes technological and economic parameters that reflect the characteristics of the technology and essentially determine the importance of adopting the new technology. Technological and economic parameters include the technology's energy efficiency, investment and operation costs, the current and highest expected implementation rate, and the technology's life expectancy. These parameters are primarily acquired through our surveys of refineries, which are supplemented with data from grey literature about production figures (e.g. Refs. [27], [28], [29] and [30]). The average values of key parameters for thirty technologies are illustrated in Table 3.

The survey results show that energy savings, efficiency, and costs for individual refineries vary in plant configuration, size, location, and operating characteristics. Therefore, in addition to the average values of key parameters (i.e. unit savings and costs), variations in plant characteristics are also recorded in the database as input parameters for the uncertainty analysis. Changes in unit savings range from 4% to 69% (average = 30%). Changes in unit costs range from 5% to 81% (average = 40%). Monte Carlo simulation was used in this study to account for the varied technology unit savings ($ue_{n,i,t}$), and the technology unit cost ($f_{n,i}$ + $oc_{n,i}$). We assumed that the changes in these values follow the uniform distribution within the given scope shown in Table 3. The Monte Carlo technique was used to produce random input values for the unit savings and unit costs (5000 times in this study). The random changes in input variables cause uncertainty in the output variables (e.g., $ES_{n,t}$, $C_{n,t}$). In this model, the output variables are linear combinations of input variables ($ue_{n,i,t}$, $f_{n,i}$ + $oc_{n,i}$), as shown in Equations (2) and (3). The distribution of output variables can be simulated based on the linear relationship. The error propagation over the output variables is analyzed by producing the distribution and determining the upper and lower bounds of a 95% confidence interval.

LIMITATIONS OF OPTIMIZATION MODEL

The optimization model describes the technology adoption on the cost-effectiveness of the financial investment. Whereas techno-economic variables are key determinants for technology choice in modeling tools, energy efficiency improvement is constrained by other factors, such as market barriers, behavioral failures and policy conflicts, as Mundaca et al. revealed [52]. This model considers the barriers by assuming a higher discount rate for the investment of energy-efficient technologies. This simple assumption lacks a detailed consideration of the barriers and their influence on technology adoption [46]. The market, behavioral and policy determinants that are not explicitly addressed in this model are identified and analyzed below.

Technology adoption is considered to be a strictly rational decision-making process and assumes a perfect market. However, information asymmetries result in a lack of knowledge about the availability of an energy efficiency technique and its characteristics, such as cost and saving potentials as well as the actual energy consumption of the equipment in place [47]. Based on a survey by De Groot et al., 30% of interviewed companies were found to be unaware of new, energy-efficient technologies [48]. Even under awareness, bounded rationality can still lead to a low/slow technology adoption. For example, our survey found that a new low-NO_x burner (WBNQ-DQ500) helped to reduce the use of fuel oil by 4.5 kg/t-feed and controlled NO_x emission below 20 ppm. Despite these advantages, most of small refineries in China are incapable to compare alternative types of burners. These refineries are inclined to replace the broken burner with a new one of the same brand and type as quickly as possible because the cost of short production interruptions are at least five times the burner price. Even larger refineries primarily choose their burners on the basis of "satisfactory" rather than "optimal" behavior [51].

The optimization model only considers the financial costs (investment costs and energy savings) but neglects transaction cost and access to the capital. These costs prevent refineries from undertaking energy efficiency projects under the label of information asymmetries. Transaction costs typically account for 3–8% of the necessary investment, of which 2–6% is attributed to information gathering, 1–2% to decision-making and less than 1% to monitoring [49]. The presence of transaction costs is likely to make potentially profitable investment unattractive and favor inefficient or generic energy technologies. In addition to significant transaction costs, prioritizing the sharing of funds for environmental compliance or product-related research with a more transparent return on investment further erodes the competitiveness of new energy efficiency projects [15]. This situation was proved by the research of Harries [50], in which 35% of such projects with attractive payback possibilities were not enacted because they were assigned a lower priority than investment projects in the core business.

This model presumed a maximum implementation rate of 90% for most of the energy-efficient technologies; however, the heterogeneity between refineries and markets has been discussed to lower the expected technology diffusion [47]. A case study conducted by de Almeida revealed split incentives between different market players in the adoption of highly efficient electric motors. A high degree of heterogeneity was observed without the dominance of a single technology, even if the technology was significantly more cost-effective than others [47].

Whereas capital and operating costs are relevant for efficient technologies, Mundaca et al. stated that they represent only a part of a great variety of determinants that drive consumers› decisions regarding energy-related technology choices [52]. Many determinants are described in terms of non-energy co-benefits, for instance, improved production efficiency, functionality, quality, reliability and design. Cost-effectiveness criterion seems insufficient to comprehensively address the attributes of energy policy instruments. In addition, corporate behavior in relation to efficient-technology choice and market failures is likely to change over time. The statistically derived trends embedded in the model›s historical data may result in unrealistic representation of decision-making when choosing technologies. Therefore, there is the need to continuously verify and validate the capability of model in relation to new data and policy considerations (e.g., dynamic efficiency, distributional equity and political feasibility).

ANALYSIS RESULTS FOR ENERGY SAVING POTENTIAL

Technology Adoption

Fig. 3 shows the technology diffusion trends for six principle refining processes. Over the next decade, the average adoption rate of energy-efficient technologies will increase from 51% to 65%. With regard to the six principle refining processes, the catalytic reformer and hydrocracker show the most significant increases in future adoption; the adoption rate for the reformer and hydrocracker will increase by 35% and 22%, respectively. The growing benefits from the rapid changes in advanced process control technology ((30) APC (Advanced process control)) is particularly suited to highly integrated operating systems associated with reformer and hydrocracker processes. In contrast, for FCC and coker processes, energy recovery technologies ((25) flare gas recovery and (27) low level heat utilization) are more likely to be adopted than APC systems because large volumes of hot flue gas are generated in these units that can be recovered in a waste heat or CO boiler to provide steam for reactor stripping or power for the compressor. For ADU,

VDU, and hydrocracker processes, upgrades to process heaters ((4) draft control and (8) new burners) are the primary adopted energy savings technology because most of the fuels supplied to these units are consumed by heating the heavy fractions.

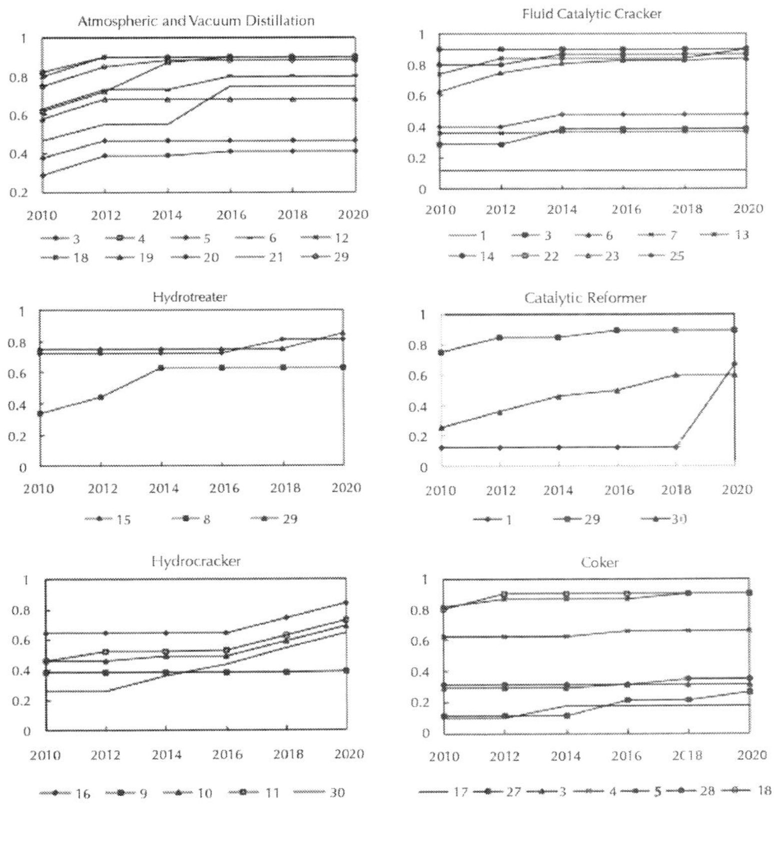

Figure 3: The changing trends for the technology implementation rates in various refining processes.

These technologies will develop at different rates. In the base year, process heaters,[2] which consume over 60% of all the fuel used in refineries, are an important area for energy efficiency improvements and will develop quickly over a short period of time. Draft control systems are a common energy-efficient technology for heaters that allow furnaces to run with 1–1.5% oxygen rather than the regular 5–6%. Draft control systems reduce energy use by 3–6%, saving refineries over 9 million RMB a year.[3] Draft control systems have been introduced in 62% of process heaters and are expected to reach an adoption rate of nearly 90% within the next five years. A similar trend is also found in the adoption of new burners. Because NO_x emission rates will require improvement in almost all refineries to meet air quality standards, new burners are being developed that aim to improve fuel and air mixing and transfer heat more efficiently. At this time, 34% of refineries have installed new burners in hydrotreater units. New burners eliminated the need for a selective catalytic reduction (SCR) and saved electricity costs associated with operating compressors and fans for the SCR. The installation of new burners is estimated to produce energy savings of approximately 20 RMB/t-feed and is expected to be implemented in 60% of refineries within five years. There are two reasons for the rapid diffusion of heater-related technology:

- First, furnaces and boilers in Chinese refineries need significant improvements. Currently, the average thermal efficiency of furnaces is estimated at approximately 75%. If unavoidable heat loss and the dew point are taken into account, the theoretical maximum for thermal efficiency is approximately 92% (HHV) [4] and [31]. This finding suggests that a 17% average improvement in energy efficiency can be achieved through changes in furnace and burner design. A 1% improvement in furnace efficiency could save a refinery over 1 million RMB. Thus, investing in heater-related technologies is quite profitable.
- Second, the technologies for heaters are relatively easy to install. Refineries prefer to adopt a new technology that has the lowest impact on the original process because this helps reduce not only the cost of reconstruction but also the risk of unsteady operation caused by process changes.

In the mid-to-long run, heat recovery[4] and advanced process control systems will gradually dominate the technological marketplace and eventually become the most important technologies in Chinese refineries. Two types of heat recovery technologies and APC systems will become increasingly important:

- The first type of important heat recovery technology involves the installation of heat pipes in air conditioning systems to enhance the efficiency of both dehumidification and energy use. This process would increase moisture removal by 30–50%. Heat pipes reduce fuel use by 120 tons, resulting in an annual cost savings of 100 thousand RMB for a refinery.
- Second, hot water tracing can be used as a substitute for steam to maintain the temperature of pipe systems. Hot water tracing reduces steam consumption by 2–4 tons per hour and operation costs by approximately 15 RMB/t-feed. The two technologies change rapidly because they have low technology and service requirements for personnel. It is relatively easy to add equipment and pipes to existing systems when process requirement demands and insulation are optional due to excess heat capacity.

In addition to heat recovery, the use of APC systems will also be widely disseminated in the future. APC systems can reduce the time required to perform complex tasks, improve production consistency, and optimize process operations based on the acquisition of more accurate data [4]. The energy savings vary from 3.9 to 21.1 MJ/t-feed depending on plant operating characteristics and the skills of system operators, which is one reason why the adoption of APC systems is delayed.

Energy Saving Potential Estimation

The potential for energy savings in petroleum refining processes can be further evaluated based on technology diffusion trends. Fig. 4 shows that the use of the energy-efficient technologies would yield 2.7×10^8 GJ in energy savings over the next decade, which is approximately 21% of the total energy consumed by this industry in 2010. In practice, areas for improvement may overlap and result in lower savings. Based on the current energy consumption rate (61 kgoe/t) and the largest potential savings, the minimum amount of energy consumed in 2020 will be 57 kgoe/t. The economic feasibility of these savings was also evaluated. The results show that the energy savings may not be sufficient to balance the cost of popularizing these technologies. Technological costs are anticipated to reach 6.7 billion RMB by 2020, which would account for approximately 9% of this industry's annual profits if profits remain at 136 RMB/t.Fig. 4 also shows the scope of uncertainty for the estimated potential energy savings and associated costs. Changes in energy savings and costs (the standard deviations for energy savings and cost) gradually increase over time, up to 2.18×10^7 GJ for energy savings and 1.5 billion RMB for costs.

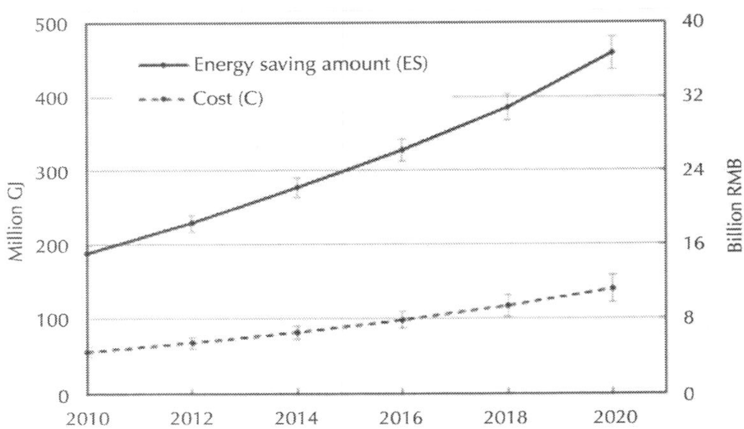

Figure 4: Changing trends in energy savings and technological costs in China's petroleum refining industry.

We have attempted to compare the results for the refining industry with existing studies on other Chinese industries. Annual potential energy savings for the refining industry are estimated at 27 million GJ, while the potential savings of the iron steel, cement, power and transportation industries are 675, 1,426, 14, and 300 million GJ, respectively [36], [37], [38] and [39]. These results are reasonable because petroleum refining and power production are nationally owned, highly efficient businesses, while iron steel and cement production are likely managed by small individual plants with increased opportunities for technology upgrades. In addition to the comparison among Chinese industries, we have also compared the Chinese results with previous studies that examined refineries in Europe, Korea, the United States, and Brazil. The potential energy savings for processing 1 t of crude oil in China is estimated at 0.32 GJ, while the potential savings in the Unites States, European countries, Korea, and Brazil are 0.53, 0.17, 0.04, and 0.11 GJ, respectively [4], [14], [17], [18] and [19]. Despite inferior technological levels, Chinese refineries have comparable potential savings because we only consider technologies that are commercially available in China. These technologies are relatively limited compared to the technologies available in the United States and Europe.

The six studied processes offer different opportunities for energy efficiency improvements (Fig. 5). The fluid catalytic cracker (FCC) showed the largest potential energy savings (up to 157 million GJ). This is unsurprising given the typically low efficiency and large capacity of the current cracking process. The technologies with the second largest potential for energy savings are atmospheric and vacuum distillation (ADU&VDU), which includes the crude desalting process, crude preheat train, fired heater, distillation column, and smaller stripping towers. The remaining processes, especially the catalytic-reforming process, are not significantly inefficient, primarily because advanced and integrated technologies are almost installed in these processes, thus providing less flexibility for future system upgrades. Considering the costs, the highest priority for improvement should be given to the coker, followed by the hydrotreater, atmospheric and vacuum distiller, and the catalytic reformer. Note that the above results are based on a simple average of the refinery's operation data. In reality, the data from different refineries may vary significantly, which may lead to a different distribution. If this uncertainty is considered, the energy savings and cost changes may reach 14 million GJ and 1 billion RMB, respectively. These changes account for approximately 14% and 24% of their respective estimated values.

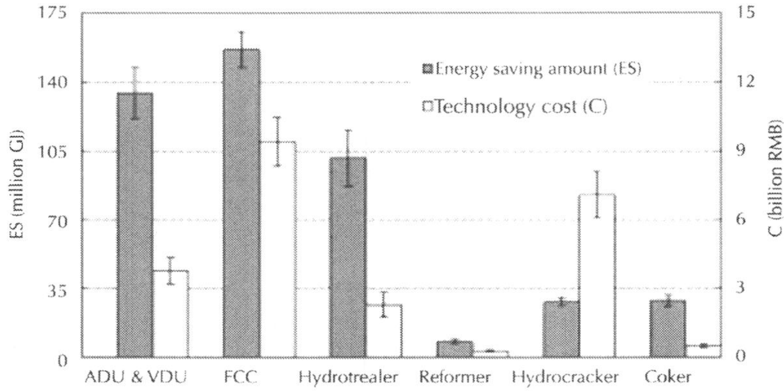

Figure 5: Estimated energy savings and technological cost by refining processes on a production per year basis.

Fig. 6 shows that process heaters and energy recovery are the major areas for refinery energy conservation in the next decade. Process heaters consume over 60% of all fuel used in refineries. Many different practices have been developed to narrow the gap in energy consumption between domestic and international advanced levels, including improvements to heat transfer and the installation of air-preheaters and soot blowers, as well as improved control. By implementing these practices, energy efficiency can be improved by an average of 10%, which is equivalent to an approximate energy savings of 10 million GJ per year for the entire industry. Energy recovery is also a strategy for the cost-effective reduction in energy consumption that maintains production quality in the refining process. The major refinery process technologies that would benefit from developments in energy recovery are the flare gas recovery system, stream generation and distribution, and the power recovery turbines. Implementation of these technologies would provide an annual savings of 8.2 million GJ, which is approximately 30% of the total savings.

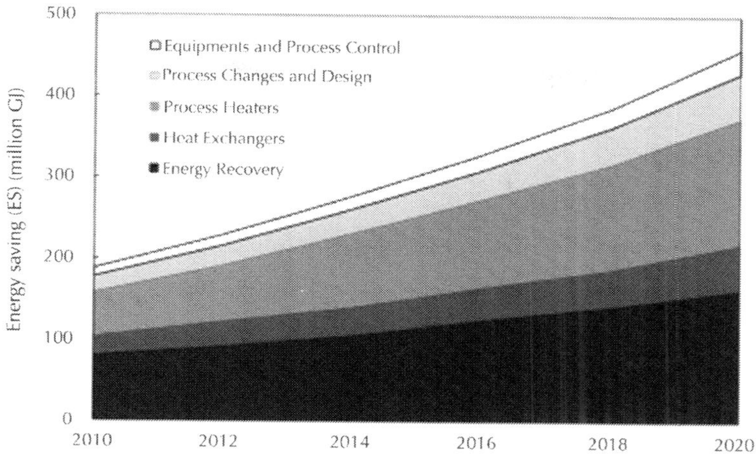

Figure 6: The major areas for energy conservation in refineries over the next decade.

Carbon Mitigation Analysis

The emission mitigation analysis indicates that implementing a cap-and-trade scheme for technology promotion is likely to further reduce accumulated energy use by 1.3×10^8 GJ over the next decade and maintain CO_2 emission levels at 87.3 Mt by 2020, lower than the designed emission cap of 93.8 Mt. Annual emissions is expected be reduced by 1.7 Mt, which is approximately two-thirds of the annual reduction of the Chinese coal-based ammonia (2.5 Mt) and caustic soda industries (2.2 Mt) and one-fifth of the Chinese calcium carbide industry's annual reduction (7.4 Mt) [42] and [43]. In the cap-and-trade scenario, the CO_2emissions of China's refining industry result in a very small decrease between 2010 and 2016, which is mainly due to the restraint of technological costs. After the CO_2 emissions price is introduced in 2016, Chinese refineries must reduce their CO_2 emissions by expanding the deployment of process heaters and energy recovery technologies. Because CO_2 emissions are strongly related to energy efficiency, process heaters and energy recovery remain the dominant technologies to achieve significant energy savings (much like in the base scenario), as shown in Table 4.

Table 4: Cost-effective and total technical potential for energy savings and CO_2 emission reduction in the Chinese refining industry by 2020

Process	Base scenario		Cap-and-trade scenario		Dominant technologies
	RMB/GJ	RMB/tCO_2e	RMB/GJ	RMB/tCO_2e	
ADU & VDU	6.2	79.6	6.4	82.9	4,5,12,18
FCC	2.7	34.8	0.7	8.6	13,22,23
Hydrotreater	4.1	52.6	4.1	52.6	15,29
Reformer	12.2	157.1	13.6	175.1	29
Hydrocracker	13.6	175.8	14.2	183.8	16
Coker	1.6	20.7	1.4	18.6	5,18

The optimization result of the base scenario model shows that Chinese refineries need to pay 6.7 billion RMB for technology upgrades between 2010 and 2020. The accumulated total cost is 12.2% higher in the cap-and-trade scenario. The change in total cost is closely related to the cost-effectiveness of different refining processes. Out of six refining processes, only two processes (FCC and Coker) are profitable for one ton of carbon mitigation under the initial carbon price (50 RMB/t). Furthermore, these processes likely minimize emissions because they enjoy low reduction costs and could benefit from selling CO_2 credits under this scheme. Carbon trading significantly improves the cost-effectiveness of emission reduction for the FCC (74.0%) and coker (12.5%), while it reduces the cost-effectiveness for the other four processes. The energy efficiency measures for the other processes are technically applicable, but they are not cost-effective under this scheme. This observation also partly proves the importance of developing low-cost, energy-efficient technologies to reduce the total cost of China's refining industry [35]. When performing sensitivity analysis, the carbon price range is extended to span from 10 RMB/t to 100 RMB/t, with a fixed implementation time and increasing annual rate. The CO_2 emitted by the refining industry is found to be lower over the whole time period than the assigned cap only when the carbon price is higher than 35 RMB/t. A higher carbon price stimulates this industry to develop large-scale process heaters and energy recovery and further decrease the emission costs (or increase mitigation revenue). When the assigned price reaches 100 RMB/t, the CO_2 emission level is expected to decrease to 72.7 Mt in 2020, 30.8% lower than the emission level in the base scenario; however, the change of carbon price mainly influences the development of energy-efficient ADU & VDU, FCC, hydrotreater and coker technologies, while reformer and hydrocracker technologies are minimally influenced because reductions in their emissions are costly.

The Impact of Barriers on the Modeling Results

As outlined above, substantial energy savings would be possible for Chinese refineries without a significant increase in capital investment if the suggested energy-efficient technologies are applied. However, although these technologies are commercially available, they are often not fully implemented due to the market barriers and policy constraints mentioned in Section 2.6.

Overall, information asymmetries and bounded rationality are likely to reduce the anticipated implementation rate of cost-effective technologies and thus reduce potential energy savings. For example, fuel-efficient catalysts for hydrogen plants are modeled to save Chinese refineries 2.3×10^7 GJ fuel oil because this technology would combat the rising price of natural gas. However, the energy savings associated with implementing this new technology are currently underestimated due to previously low fuel costs, which may lead to lost opportunities for the diffusion of fuel-efficient catalysts. Furthermore, the absence of updated codes and standards for new technologies reduces the chances of adoption. For example, plant owners may not be convinced that a shock wave soot blower (technology (7)) is sufficiently reliable and thus would prefer to continue using the conventional soot blower.

A lack of access to capital and burdensome transaction costs may further hamper the diffusion of remarkable energy saving technologies with prohibitively initial installation investment. For example, the heat pipe technique (technology (21)) is modeled to increase implementation rate by 28% and save 1.2×10^7 GJ for ADU&CDU. For an individual refinery with an annual capacity of 3.8 million tons, the initial installation investment would be 5 million RMB, which implies that the search and information gathering in an imperfect market would cost approximately 250 thousand RMB more [49]. Significant transaction costs make this technique seem more expensive than conventional ones. This situation is aggravated due to the existence of the heterogeneity between refineries and markets, which may limit the maximum implementation of relatively expensive dominant technologies (e.g., higher efficiency feed nozzles and heat high-pressure separating processes) and become favorable for the diffusion of inferior energy-efficient technologies (e.g., distributed control systems and water reuse).

As a whole, the modeling results may underestimate the costs of reducing energy demand and overestimate the potential of technological change to improve energy efficiency if the above market and behavioral barriers are not considered. The deviations would be partly compensated by effects of the experience curve. The use of experience curves (a learning rate of 20% is suggested) offers future capital cost savings relative to cumulative production and increases the opportunities of technology diffusion through network-building and use-cost reductions [52]. Therefore, in light of complex market environment and inevitable policy trade-offs, cost-effectiveness is unlikely to be optimized on a continual basis, and the regulators may choose a path that may seem suboptimal in the long term from the model perspective. Results from ex post evaluation would be central to complement and generate better insights that further advance the modeling tools, as suggested by Mundaca et al. [52].

CONCLUSIONS

To meet the energy challenges faced by Chinese petroleum refineries, a methodology was developed to investigate the potential energy savings new technologies could provide for this industry. This method included a comprehensive tool for decision-making based on a bottom-up model of production operations and technology diffusion theory. Based on the constructed model, the potential energy reduction offered by 30 energy-efficient technologies applicable to the six examined processes were evaluated and compared to provide recommendations for energy conservation strategies in this sector. The results not only provide insight into the feasibility of certain energy reduction goals based on available technology, but they also provide reference information for gradual technological policy design. The primary findings and recommendations from this research are twofold:

First, process heater upgrades are identified as a priority to enable short-term energy optimization and control in refineries. With increasingly stringent fuel formulation requirements and energy conservation objectives, heat recovery and advanced process control will gradually occupy the technological marketplace and eventually become the dominant technology in Chinese refineries. Two key barriers should be overcome to further improve plant efficiency and profitability: 1) New, previously unconsidered technologies, such as improved catalysts, should be re-evaluated in light of higher gas prices; and 2) codes and standards need to be updated to demonstrate the reliability of new technologies, such as shock wave soot blowers.

Second, technology diffusion driven by current policies will control energy consumption within 57 kgoe/t in the next decade. Through the establishment of cleaner production standards and the diffusion of energy technologies, a maximum of 2.7×10^8 GJ can be saved; this value is approximately 21% of this industry's total energy consumption in 2010. It should be noted that the potential savings generated by the model are based on social planning with perfect knowledge, which implicitly assumes rational decision-making and a perfect market. In reality, policy rationality may stress effectiveness, equity issues, timing and risk, and therefore result in a path that is not cost-optimal in the long term. Although the uncertainty of real practice is somewhat tempered by higher discounts, the model may still tend to be optimistic in terms of efficiency gains due to its implicit technology optimism.

This paper is designed to assess key energy-saving technologies in Chinese refineries, with an emphasis on their technical potential. The results may underestimate the potential savings for the Chinese refining industry as a whole for two reasons: 1) not all available energy-efficient technologies are assessed in the present study; 2) our estimation reflects only the technical potential for energy efficiency. Fuel substitution (i.e. replacing the liquid fuel with natural gas) could also reduce energy consumption, an option that was not considered in our study. For a more thorough estimation for the whole industry, all energy-saving options should be investigated in more detail.

Nevertheless, our results indicate areas with the potential to adopt thirty commercially available technologies. Furthermore, our results show the relative magnitude of the energy-saving potential from using such technologies under various policy scenarios. The results further help policy-makers understand industry needs and areas where technologies will have maximum impact. This understanding will allow the extension of potential strategies from the current energy efficiency policy to a more flexible cap-and-trade-like policy. The carbon cap and price mechanism is expected to avoid approximately 4.0×10^8 GJ in energy and 17 Mt in carbon emissions, which are very sensitive to the carbon trading price. The carbon cap and price mechanism influences the cost-effectiveness of various refining process quite differently. Both the FCC and coker benefit from emission trading; however, significant emission reductions in the other processes may require financial incentives beyond carbon price alone.

Increased air quality and climate change demands coupled with new developments in automotive technology and biotechnology pose new challenges for Chinese refineries. These refineries must balance the competing goals of profitability, energy efficiency, and environmental compliance with increased demands for product quality. Energy efficiency improvement through technological retrofitting is a relatively low-cost way for refineries to reduce emissions and to cut operating costs. To meet future energy demands, energy efficiency improvements must be combined with technological research and development that utilizes alternative energy sources and on-site power generation. This process will require not only significant administrative effort but also innovative partnerships among refineries, utilities, and non-utilities.

ACKNOWLEDGMENTS

This research was financially supported by the Chinese Ministry of Science and Technology (MOST, Project No. 2009BAC65B03). The authors are very grateful to the two reviewers for their constructive suggestions and for their proposed corrections to improve the paper.

REFERENCES

1. Koottungal L. 2009 worldwide refining survey. Oil Gas J. Available at: http:// www.cne.es/cgi-bin/BRSCGI.exe?CMD¼VEROBJ&MLKOB¼475026700505; December 2009 [accessed 05.05.13].
2. China Petroleum and Chemical Industry Association. Yearbook of China chemical industry 2009. Beijing: Center of China Chemical Industry Information Press; 2010.
3. National Bureau of Statistics of China. China statistical yearbook 2010. Beijing: China Statistics Press; 2010.
4. Worrell E, Galistsky C. Energy efficiency improvement and cost saving opportunities for petroleum refineries: a ENERGY STAR guide for energy and plant managers. Ernest Orlando Lawrence Berkeley National Laboratory; February 2005.

5. Meng XL. A review of energy saving in China's refining industry. Pet Petrochem Today 2005;13(3):31-4.
6. China Petrochemical Consulting Corporation. Yearbook of China petrochemical consulting corporation 2009. Beijing: China Petrochemical Press; 2009.
7. China's Ministry of Environmental Protection. Cleaner production standard: petroleum refinery industry (semi-asphaltic flux) HJ443-2008. Beijing: China Environmental Science Press; 2008.
8. China National Development and Reform Commission (NDRC). Category of key technology for energy efficiency improvement (third batch). Beijing: NDRC; 2010.
9. Bai Y. Restructure and development of China petroleum and chemical industry during the twelfth five-year plan period. Chem Ind 2010;6:4-10.
10. Xu HF. The global refining industry: development and trends. Int Pet Econ 2009;17(5):1-10.
11. Zhang RY. Review of China refining industry operation in 2009 and outlook on the next two years. Int Pet Econ 2010;5:22-6.
12. Walls WD. Petroleum refining industry in China. Energy Policy 2010;38: 2110-5.
13. Gueddar T, Dua V. Novel model reduction techniques for refinery-wide energy optimization. Appl Energy 2012;89:117-26.
14. Energetics Incorporated for the U.S. Department of Energy. Energy bandwidth for petroleum refining processes. Maryland: Industrial Technologies Program (ITP); 2006.
15. Energetics, Incorporated Columbia. Energy efficiency roadmap for petroleum refineries in California. California: Energy Commission; 2004.
16. ICF International. Energy trends in selected manufacturing sectors: opportunities and challenges for environmentally preferable energy outcomes. Philadelphia: U.S. Environmental Protection Agency; 2007.

17. Szklo A, Schaeffer R. Fuel specification, energy consumption and CO2 emission in oil refineries. Energy 2007;32:1075-92.
18. Johansson D, Rootzen J, Berntsson T, Johnsson F. Assessment of strategies for CO2 abatement in the European petroleum refining industry. Energy 2012;42: 375-86.
19. Park S, Lee S, Jeong JS, Song HJ, Park JW. Assessment of CO2 emissions and its reduction potential in the Korean petroleum refining industry using energyenvironment models. Energy 2010;35:2419-29.
20. Brechet T, Tulkens H, Beyond BAT. Selecting optimal combinations of available techniques, with an example from the limestone industry. J Environ Manag 2009;90:1790-801.
21. Rafiqul I, Weber C, Lehmann B, Voss A. Energy efficiency improvements in ammonia production-perspectives and uncertainties. Energy 2005;30: 2487-504.
22. Kuckshinrichs W, Zapp P, Poganietz WR. CO2 emissions of global metalindustries: the case of copper. Appl Energy 2007;84:842-52.
23. Zhang C, Wen ZG, Chen JN. An integrated model for technology forecasting to reduce pollutant emission in China's pulp industry. Resour Conserv Recycl 2009;54:62-72.
24. Xu JL. The potential and countermeasures of saving energy in oil refining enterprises. China Foreign Energy 2006;11:8-12.
25. Qi XF. Technology diffusion model and policy implications. Jianghan Forum 2004;6:5-9.
26. Zhang DM, Li YJ, Tu QH. Status of China petrochemical industry in 2010 and development forecast in 2011. China chemical industry annual report in 2010. China: Chemical Industry; 2011.
27. Lin SX. Petroleum refining engineering. 4th ed. Beijing: Petroleum Industry Press; 2010.
28. Wang WT. Case study of energy-saving reform of petrochemical process. Beijing: Chemical Industry Press; 2008.

29. Hua G. Study on optimum technology of energy system optimization in petroleum refiners. Beijing: China Petrochemical Press; 2009.
30. Xu CM, Bao XJ. Technology progress in petrochemical industry. Beijing: Petroleum Industry Press; 2006.
31. Petrick M, Pellegrino J. The potential for reducing energy utilization in the refining industry. report nr. ANL/ESD/TM-158. Lemont: Argonne National Laboratory; 1999.
32. Jaccard M, Nyboer J, Sandownik B. The cost of climate policy. Vancouver: UBC Press; 2002. p. 175-98.
33. Rehan R, Nehdi M. Carbon dioxide emissions and climate change: policy
34. National People's Congress. The outline of the national 12th five-year plan on economic and social development. Beijing: Renmin Press; 2011.
35. Dai Y, Hu X, Jiang K, Xu H, Zhu Y, Bai Q, et al. Low carbon development pathway for China towards 2050-scenario analysis on energy demands and carbon emissions. Beijing: Science and Technology Press; 2009.
36. Hasanbeigi A, Morrow W, Sathaye J, Masanet E, Xu T. A bottom-up model to estimate the energy efficiency improvement and CO2 emission reduction potentials in the Chinese iron and steel industry. Energy 2013;50:315-25.
37. Ke J, Zheng N, Fridley D, Price L, Zhou N. Potential energy savings and CO2 emissions reduction of China's cement industry. Energy Policy 2012;45: 739-51.
38. Lin B, Wu Y, Zhang L. Electricity saving potential of the power generation industry in China. Energy 2012;40:307-16.
39. Yan X, Crookes RJ. Reduction potentials of energy demand and GHG emissions in China's road transport sector. Energy Policy 2009;37:658-68.
40. Herold A. Comparison of CO2 emission factors for fuels used in greenhouse gas inventories and consequences for monitoring and reporting under the EC emissions trading

scheme. Netherlands: European Topic Centre on Air and Climate Change; 2003.

41. The Climate Registry. U.S. default factors for calculating CO2 emissions from fossil fuel and biomass combustion. Available at: http://www.theclimateregistry.org/downloads/2012/01/2012-Climate-Registry-Default-Emissions-Factors.pdf. [accessed 23.02.13].

42. Zhu B, Zhou W, Hu S, Li Q, Griffy-Brown C, Jin Y. CO2 emissions and reduction potential in China's chemical industry. Energy 2010;35:4663-70.

43. Liu X, Zhu B, Zhou W, Hu S, Chen D, Griffy-Brown C. CO2 emissions in calcium carbide industry: an analysis of China's mitigation potential. Int J Greenh Gas Con 2011;5:1240-9.

44. Lee CF, Lin SJ, Lewis C. Analysis of the impacts of combining carbon taxation and emission trading on different industry sectors. Energy Policy 2008;36:722-9.

45. Hiraishi T, Krug T, Gytarsky M, Irving W, Penman J. IPCC guidelines for national greenhouse gas inventories. In: Eggleston S, Buendia L, Miwa K, Ngara T, Tanabe K, editors. IPCC methodology report. Geneva: Intergovernmental Panel on Climate Change; 2006.

46. Worrell E, Ramesohl S, Boyd G. Advances in energy forecasting models based on engineering economics. Annu Rev Environ Resour 2004;29:345-81.

47. Fleiter T, Worrell E, Eichhammer W. Barriers to energy efficiency in industrial bottom-up energy demand modelsea review. Renew Sust Energ Rev 2011;15: 3099-111.

48. de Groot HLF, Verhoef ET, Nijkamp P. Energy saving by firms: decision making, barriers and policies. Energ Econ 2001;23(6):717-40.

49. Hein LG, Blok K. Transaction costs of energy efficiency improvements. In: conference proceedings: European Council for an energy efficient economy (eceee); 1995.

50. Harris J. Investment in energy efficiency: a survey of Australian firms. Energy Policy 2000;28(12):867-76.
51. de Almeida E. Energy efficiency and the limits of market forces: the example of the electric motor market in France. Energy Policy 1998;26(8):643-53.
52. Mundaca L, Neij L, Worrell E, McNeil M. Evaluating energy efficiency policies with energy-economy models. Annu Rev Environ Resour 2010;35:305-44.

Chapter 7

Carbon Dioxide (CO_2) Estimation from Kuwait's Petroleum Refineries

S.M. Al-Salem

Petroleum Research Center, Kuwait Institute for Scientific Research, Safat 13109, Kuwait

ABSTRACT

A detailed study of carbon dioxide (CO_2) sources from petroleum refineries in Kuwait was conducted. The aim was to estimate the amount of CO_2 being emitted and determine the influencing factors on physical and chemical process in refineries. Three refineries were considered in this work, processing around 900 Mbpd of crude oil. Process heaters stacks constituted 62–75% of the CO_2 emissions, ranking at the top of the sources. Hydrogen production (HP) units via steam-methane reforming ranked seconded in the final assessment, ranging from 12% to 25% in the studied refineries. The configuration of the refinery was determined to be a critical factor in such work. The properties of the processed oil, in terms of density, sulfur content and API were essential in comparing results obtained in this work to other refineries specific rates. Carbon capture and storage (CCS) presents itself as a viable solution for the reduction of the carbon footprint of Kuwait downstream sector, especially when considering the purity of the HP units product gas. The work presented in this paper can be considered as the cornerstone of future carbon mitigation plans in the country.

INTRODUCTION

One of the main contributing factors to CO_2 emissions is the demand on low sulfur clean fuel (e.g. diesel, gasoline, etc.), which increases carbon emission intense processes in refineries. This forces petroleum refineries to start implementing CO_2 mitigation measures in their practices. Carbon capture and storage (CCS) is considered to be a promising solution for further CO_2 reduction in petroleum refineries. However, the use of CCS is so far focused on power plants since they are the major CO_2 contributor to the world stationary sources (IPCC, 2005). Refineries present themselves as an advantageous option for CCS due to the abundance of carbon intense processes, especially units' utilities and hydrogen production (HP) units. Reports show that up to date, 74 CCS projects are active or in final developmental stages (Johansson et al., 2013). However, to integrate a carbon capture plant in a petroleum refinery the carbon sources have to be identified and the CO_2 emission from each process has to be estimated.

In 2008, it was estimated that CO_2 emissions from the European union (EU) were about 3780 million tons (≈75% of total greenhouse gases – GHGs – emissions). Refineries in the EU account for about 8% of total CO_2 emissions in Europe. Emissions from EU refineries have increased by 17% compared to 1990 levels. This is due to the increase in demand on transport fuel (e.g. diesel) (Johansson et al., 2012). The European Commission (EC) has implemented an emission trading system (ETS) with a roadmap until the year 2050 to incrementally reduce industrial carbon emissions (EC,

2014). Since the year 2005, EU refineries are a part of the EU ETS which aim at reducing GHGs by 21% in 2020 relative to the 2005 levels (Johansson et al., 2012). In the UK, an 80% reduction target (based on 1990 levels) for GHGs was set by the government (CA, 2008). Many developing countries have started to consider CCS as a viable option for carbon emission reduction. Quantifying carbon emissions and their contribution to the total refinery carbon load is the first step in implementing carbon reduction strategies and developing CCS projects in the near future.

Kuwait produces 3 million bpd of crude oil and the Kuwait National Petroleum Company (KNPC) is responsible for refining and marketing petroleum products within Kuwait and currently operates three refineries in the country. These are: Mina Al-Ahmadi (MAA) with a processing capacity of 466 Mbpd, Shuiba (SHU) with a capacity of 200 Mbpd and Mina Abdullah (MAB) with a capacity of 270 Mbpd (Al-Ajmi, 2010). The fourth refinery (Al-Zour) will be commissioned by 2018 with an approximate capacity of 600 Mbpd. However, no CO_2 assessment is available for these refineries in literature and the routes of CCS cannot be investigated without an assessment detailing current process emissions. A past study conducted in Kuwait for major CO_2 sources was published (JNOC, 1998). Steam methane reforming (hydrogen production units) and acid gas removal (AGR) were concluded to be the major sources of CO_2 in the downstream industry, with available CO_2 in the amounts of 28–43 million SFCD and 5 million SCFD, respectively. Past research in Kuwait also included CO_2 recovery options for crude oil from power stations (Salman and Oskui, 2004), as well as an emission inventory for primary pollutants in/from the downstream industry (excluding CO_2) (Kam et al., 2005, Chehadeh et al., 2004 and Al-Salem and Khan, 2010).

CO_2 sources in a petroleum refinery could be divided to (Stockle and Bullen, 2008, van Straelen et al., 2009, van Straelen et al., 2010 and Johansson et al., 2012):

- Units' utilities (heaters, furnaces and boilers) which can contribute up to 65% of the total CO_2 load of the facility.
- Fluid catalytic cracking (FCC) unit which can be responsible

to up to 40% of total carbon emissions from the regenerator's stack due to catalyst coke burn off.

- Hydrogen production (HP) which produces CO_2 as a byproduct of the gas/water shift reaction and can emit up to 20% of the total refinery's CO_2 emission.
- Utilities production if a refinery produces steam or electricity. CO_2 is a direct by-product of such process which can contribute to 50% of the total refinery CO_2 emission load.

The USEPA (2010) stated that GHG emissions from petroleum refineries were estimated to be 214 million tons of CO_2 equivalents based on production rates in 2005. In addition, the International Panel on Climate Change ranks petroleum refineries third among world stationary sources with an estimated combined emission rate of 798 million tpa (IPCC, 2005). Other CO_2 emitters include the power sector (10,539 million tpa) and cement industry (932 million tpa). In light of the aforementioned and considering that Kuwait is an oil dependent country, a study of its major CO_2 sources in its current operating refineries is considered as a priority. In addition, analysis of this work's results is of great importance for future mitigation strategies in the petroleum sector, concerning carbon footprint reduction. This work details the major sources of CO_2 in the three refineries currently operating in Kuwait, showing the estimation methodology followed and main influencing factors on the CO_2 emission rate.

CO_2 ESTIMATION METHODOLOGY

Three refineries operate in Kuwait, processing various types of crude oils with an API around 30. Heavier crudes are also processed in Kuwait (e.g. Eocene, lower Faris, etc.) characterized with high sulfur content, hence requiring a higher demand of hydrogen (H_2) for hydrotreating and more intense energy requirements for direct heating. Electricity is imported to the three refineries from the national grid and a trade-off exists between the oil sector and the Ministry of Electricity and Water (MEW). In this section, the methodology for CO_2 estimation is detailed for each unit with a description of the process. Onsite and industrial data were gathered from each refinery to produce a comprehensive assessment of carbon sources from each refinery's facilities.

Hydrogen Production (HP) Units

Hydrogen production (HP) units are available in all three refineries in Kuwait. In MAA, there are four steam methane reformers (SMR) dedicated for hydrogen gas production. They are of an identical design capacity producing (49.5 MMSCFD hydrogen production) of 97% purity. Utilization capacity of each unit varies between 55% and 76% according to the year 2012 average data. The main function of the units is to meet the hydrogen demand of the refinery, and in particular the hydrocracking (HCR), atmospheric residue desulfurization (ARDS), kerosene desulfurization (KD) and gas oil desulfurization (GOD) units.

The CO_2 produced in the shift reaction is removed by a Benfield solution system (K_2CO_3 + diethanolamine (DEA) + H_2O). Fig. 1 illustrates the hydrogen production (HP) scheme via steam methane reforming (SMR) in the current refineries. The feed gas is desulfurized by absorption to remove all sulfur compounds using a di-isopropanolamine (ADIP) solution. A high temperature shift converter is used to convert the bulk of the CO to CO_2. A Benfield solution is used to remove the CO_2 produced from the shift converter. The solution is then regenerated and the CO_2 is vented to atmosphere. The residual gas containing CO and CO_2 is subjected to methanation, converting them to methane.

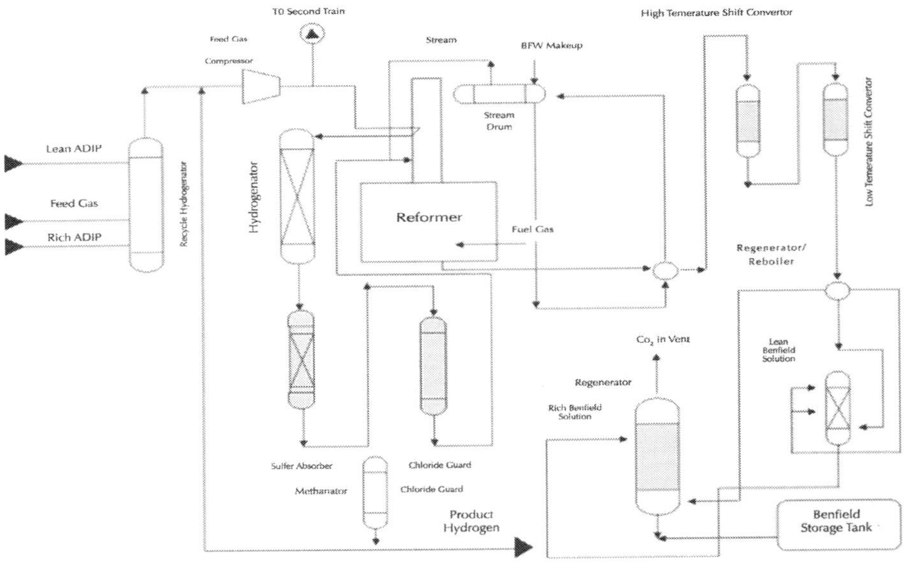

Figure 1: Hydrogen production (HP) in the current refineries schematic diagram.

The utilization capacity was multiplied by each hydrogen production rate to estimate the actual daily hydrogen production. From the stoichiometric ratio of CO_2 to H_2 from the balanced chemical reactions depicted in Eqs. (1) and (2), we can obtain the ratio of CO_2 to hydrogen, which is 1–4. Hence the CO_2production on a daily basis can be easily calculated using Eq. (3):

$$C_nH_{2n+2} + n\,H_2O \leftrightarrow n\,CO + (2n+1)\,H_2$$

$$\text{–For saturated hydrocarbons} \quad (1)$$

$$CO + H_2O \leftrightarrow CO_2 + 2H_2$$

$$\text{–Water/gas shift reaction for } H_2 \text{ recovery.} \quad (2)$$

$$HPC = HPR \cdot UC \cdot \left(1/4\right) \quad (3)$$

where HPC is the total carbon dioxide (CO_2) emission rate from the HP unit (MMSCFD); UC is the HP unit utilization capacity (%)

and HPR is the hydrogen production rate of the unit (MMSCFD). Using the conversion factor for CO_2 which is 17,483 SCF/ton, one can convert the CO_2 production to a unit mass (LG, 2011). The total CO_2 produced from HP units in MAA refinery was estimated at 7.1 × 10^5 tpa without considering the heating duty of the unit (indirect burden).

Similar to the case of MAA, SHU produces hydrogen via three identical HP units of 74 MMSCFD. The former consists mainly of hydrocarbons, while the latter is a mixture of hydrogen rich bleed gas. The design is to meet a 95% hydrogen purity with a CO_2 content of 10 ppm. The C_1, C_2, C_3 and C_4 contents in the feed are (mol%): 39.96, 41.68, 13.25 and 4.17, respectively. The process is similar to the one described previously in MAA. The utilized capacity of the HP units in SHU ranges between 58% and 69%, which was used in estimating the CO_2 emissions from these units following the same methodology described earlier for MAA. This results in a total CO_2 emission from HP units in SHU of 7.3 × 10^5 tpa. Furthermore, MAB has four HP units. Three are identical parallel trains of 55 MMSCFD capacities (gross hydrogen) with 97% minimum purity. The CO + CO_2 concentrations in the product are of 20 ppm. The fourth unit reforms desulfurized gas which is treated in this unit to meet the hydrogen demand of the atmospheric residue desulfurization (ARDS) unit. The unit is of a 39.5 MMSCFD capacity. The utilized capacity of the HP units in MAB ranges between 53% and 73%. This results in a total CO_2 emission from HP units in MAB of 7.2 × 10^5 tpa. CO_2 emissions from HP units in Kuwait could be summarized as shown in Table 1. A drastic change was estimated in the emissions of CO_2 from HP from 1998 levels (JNOC, 1998). This could be attributed to the number of revamps the refineries has faced in the past decade, in addition to, the increasing demand of H_2 in Kuwait over the past 5 years and the change of feed gas properties with time.Fig. 2 shows the distribution of HP units emission with respect to each refinery assessed. There is an even distribution between the three refineries when it comes to CO_2 emissions from HP units. The capacity of each refinery differs, however the HP utilization capacity is almost similar in each refinery. And the units in SHU and MAB are of almost similar capacity. Hence, the minor differences in the calculations come from the purity of the product stream and the units' capacity. In addition, the demand of hydrogen gas in each refinery and the units it supplies plays a major factor in the amount of hydrogen produced, consequently the CO_2 emitted.

Table 1: HP units CO_2 emissions in Kuwait from the three existing refineries

Refinery	Number of HP units	CO_2 emission (million tpa)	CO_2 emission (MMSCFD)
MAA	4	0.71	12,412
SHU	3	0.73	12,762
MAB	4	0.72	12,587

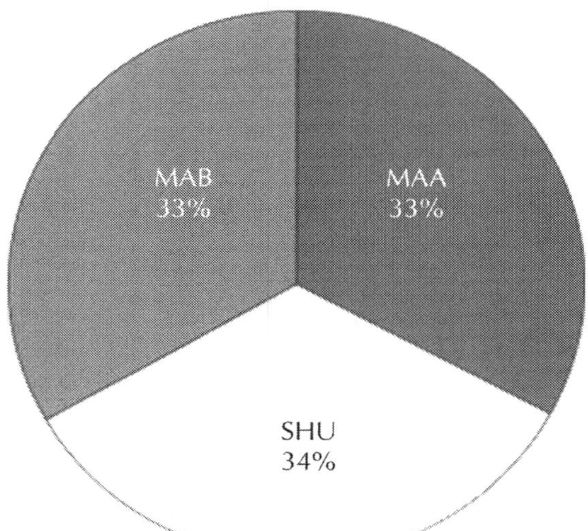

Figure 2: Percentile distribution of the carbon dioxide (CO_2) emissions from the hydrogen production (HP) units in the three refineries in Kuwait.

Fluid Catalytic Cracking (FCC)

Due to current high energy cost and progressively tightening environmental regulations, refiners have recently been expanding their FCC focus to include improved energy efficiency and reduced CO_2emissions from coke deposition on catalyst after regeneration. The main objective of fluid catalytic cracking (FCC) is the conversion of feedstock to gasoline and other valuable products. FCC also minimizes the production of less valuable products (i.e. coke, clarified oil, etc.) (Wolschlag et al., 2009). The FCC unit in MAA is the only one operated in Kuwait (Fig. 3). It has an actual operating of 43 Mbpd and processes feedstock to produce a number of products, mainly gasoline and liquefied petroleum gas (LPG). The olefin rich LPG is further processed to produce propylene, alkylate and MTBE. The reactor/regenerator section of the unit is where the cracking reactions, catalyst separation and regeneration take place. The feed to the unit is preheated raw oil (220–265 °C) combined with the hot stream of the regenerated catalyst introduced to the bottom of the reactor›s riser. The reactor vapor is sent to the main fractionator to be separated to heavy and light cycle oils, distillate, heavy gasoline and overhead vapors. Eq. (4) can be used to estimate the production of coke (by weight) and consequently CO_2 in the FCC based on the volumetric capacity of the FCC unit.

$$CFP = 0.158\,OC \cdot cf \cdot SG \cdot UC \tag{4}$$

where CFP is the coke production rate (kg/day); OC is the unit operational capacity (bpd) converted to m^3/day using the conversion factor of 0.158 m^3/bbl for a crude oil barrel; cf is the coke weight fraction in the final FCC product (4.62%); UC is the utilization capacity (89.2%); SG is the specific gravity of crude oil at 60 °F (900 kg/m^3). This results in a coke production rate of 252 ton/day, and by multiplying this with the ratio of molecular weights of CO_2 (44 g/mol) to carbon (12 g/mol), the CO_2 emission from FCC in MAA was estimated to be 3.4×10^5 tpa. It is also worth mentioning that CO_2 composition in FCC is estimated to be 15.2% (Al-Qattan, 2014), which is within expected range (De Mello et al., 2009). This indicates that the operation is within desired capability and CO_2

emissions if desired to be reduced should be controlled via energy optimization of heat in FCC unit, since no direct heating is supplied to such processes.

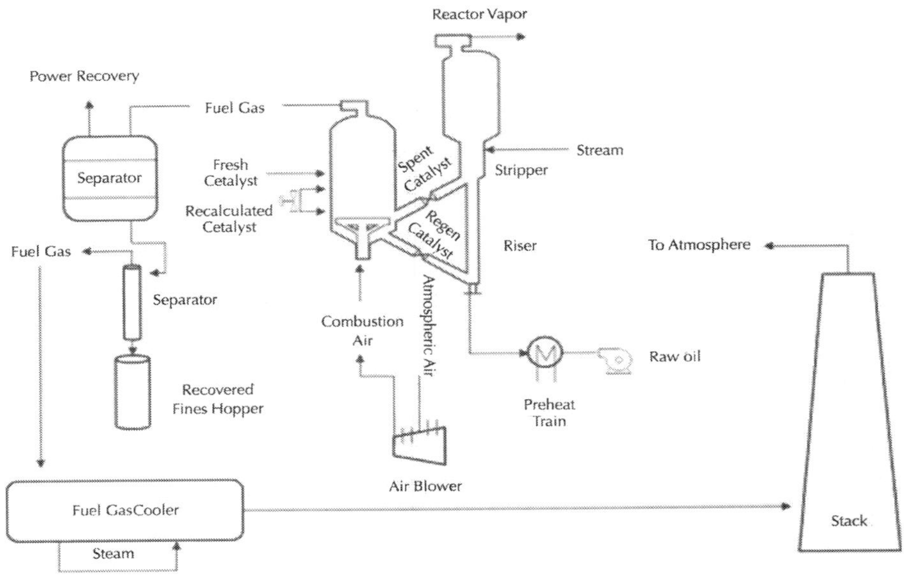

Figure 3: Flow diagram of fluid catalytic cracking (FCC) unit in studied refinery.

Fired Heaters Carbon Emission from Refineries in Kuwait

Utilities play a major role in petroleum refineries through process heaters. They supply through fired heater and boilers the energy needed for the various units to operate at desired conditions. Fuel gas is typically employed as a fuel for various utility operations. Fired heaters exhaust gas typically contain 3–12% of CO_2depending on the type of fuel being combusted (Romano et al., 2013). Other gases include excess O_2, SO_2, SO_3, N_2 and NOx. Fuel gas is used in Kuwait's refineries for fired heaters. Based on the US EPA assessment of combustion emission sources, emission factors for natural gas, fuel oil and liquefied petroleum gas (LPG) were reported as 50.6 kg CO_2/GJ, 68.6 kg CO_2/GJ and 58.7 kg CO_2/GJ, respectively. The emission factor for fuel oil was used in this work. Refinery gas emission factors reported in literature are typically refinery specific. Hence, these factors were not used and the standard fuel oil factor was considered in carbon emission estimations in this study.

MAA is the largest refinery in Kuwait with the most complex network of operation, and has two sides on its industrial location. A refinery and a gas plant side. There are 45 fired heaters in operation in the MAA refinery side. These could be classed according to its dedicated unit of service as per the following: Atmospheric distillation, vacuum distillation, hydrocracking, catalytic reforming, hydrogen generation, residual desulfurization, kerosene desulfurization and treating, distillate desulfurization and treating and others. Table 2 shows the fired heater duties and collective units' throughputs in MAA refinery. Table 3shows estimated CO_2 emission from the utilities of the refinery side. The gas plant side contains 9 fired heaters for the different LPG trains (deethanizer, depropanizer and debutanizer). Based on similar category for the past utility calculation for fuel oil, Table 4 summarizes the gas plant utility duties (reboilers) and reports the estimated CO_2 emission based on 73.96 kg CO_2/MMBTU emission factor (FDEP, 2014).

Table 2: Combined fired heaters duty in MAA categorized based on process type

Process unit type	Throughput (bpsd)	Fired heaters duty (GJ/year)
Atmospheric crude distillation	4.3×10^5	8.5×10^6
Vacuum distillation (vacuum residue + bitumen)	7.4×10^4	1.6×10^6
Hydrocracking	1.3×10^5	2×10^6
Catalytic reforming	2×10^5	4.3×10^6
Hydrogen generation	3.7×10^3 (million tons/h)	6×10^6

Residual desulfurization	7.2×10^5	3×10^6
Kerosene desulfurization & treating	2×10^4	0.1×10^6
Distillate desulfurization & treating	9.7×10^4	9×10^5
Others	7.78×10^2 (million tons/h)	1.2×10^7

Table 3: Fired heater (utilities) in the refinery side and CO_2 emissions based on fuel oil emission factor for MAA

Process	**CO_2 emission (tpa)**
Atmospheric crude distillation	5.8×10^5
Vacuum distillation	1.1×10^5
Hydrocracking	1.4×10^5
Catalytic reforming	2.9×10^5
Hydrogen generation	4.1×10^5
Residual desulfurization	2×10^5
Kerosene desulfurization & treating	1.2×10^4
Distillate desulfurization & treating	6.1×10^4
Other	8.5×10^5

Table 4: Fired heater (utilities) in the gas plant side and CO_2 emissions based on fuel oil emission factor for MAA

Fired heater	**Function**	**Duty (MMBTU/h)**	**CO_2 emission (kg/year)**
LPG train 1	De-ethanizer reboiler	186	120×10^6
	De-propanizer reboiler	163	105×10^5
	De-butanizer reboiler	105	67×10^5
LPG train 2	De-ethanizer reboiler	186	120×10^5
	De-propanizer reboiler	163	105×10^5
	De-butanizer reboiler	105	67×10^5

LPG train 3	De-ethanizer reboiler	186	120×10^5
	De-propanizer reboiler	163	105×10^5
	De-butanizer reboiler	105	67×10^5

Fig. 4 shows the CO_2 emission from MAA refinery heaters. Majority of emissions result from the gas plant side, which is quite intensive in operation for acid gas removal processing (described in Section 2.5). Activities in the gas plant include acid gas, condensate and tail gas treatment. On the refinery side, distillation units (of which there are four) process the various crudes coming into the refinery. Topping units seem to have the majority of the emissions share in the refinery side (16%, Fig. 5). It is essential to optimize the heating requirements of the topping units in order to have a better environmental friendly operation in MAA. HP units show a combined percentage of 12%. The type of feed to such units demands high utility and it has an essential demand on its heaters. The others category include all units and support units in MAA. There is vast room for improvement when it comes to the duties of these units once optimizing the control systems is conducted. Comparatively, SHU emits a total of 2.45×10^6 tpa of CO_2 via its heaters stacks; which is of a similar order of magnitude to MAA (3.55×10^6 tpa). SHU has a lower number of heaters in its industrial site that result in a high emission rate of CO_2. This is due to its intense operation in HP units, distillation and hydrocracking activities. Table 5 shows the heaters duties in SHU based on average 2012 values classed according to process type.

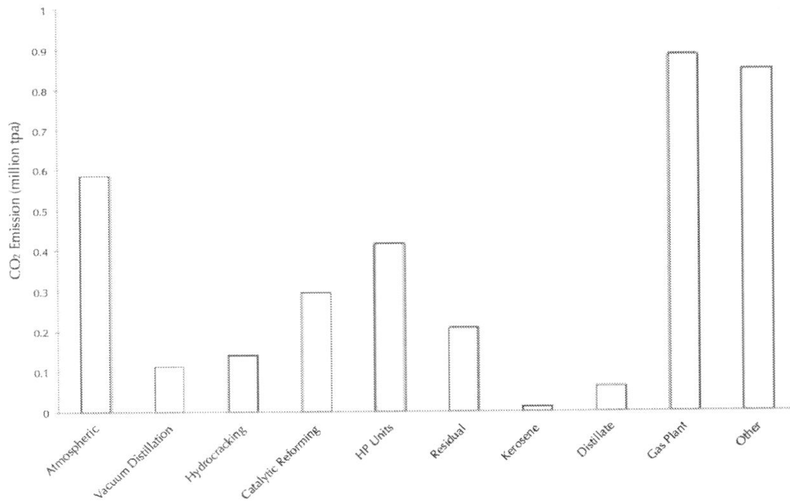

Figure 4: CO_2 emission from heaters and boilers in MAA refinery categorized per combined units section.

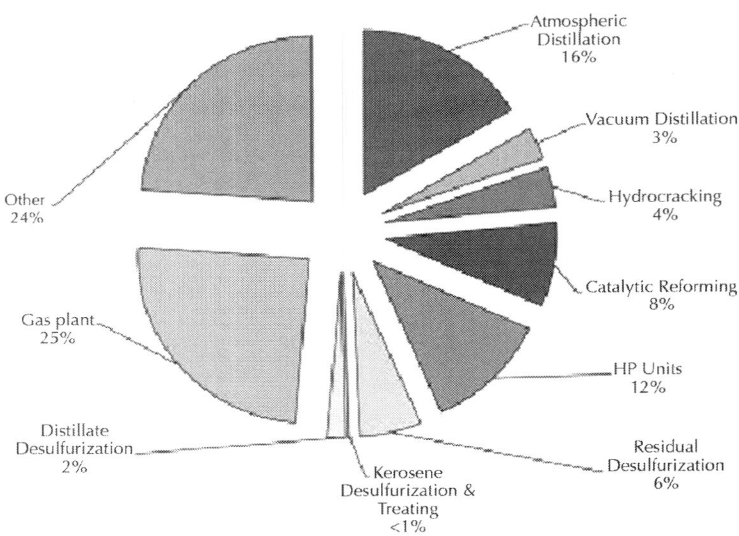

Figure 5: Percentile distribution of CO_2 emission from heaters and boilers in MAA.

Table 5: Combined fired heaters duty in SHU categorized based on process type

Process unit type	Throughput (bpsd)	Fired heaters duty (GJ/year)
Atmospheric crude distillation	3.5×10^5	4.8×10^6
Vacuum distillation	18.1×10^4	1.5×10^6
Hydrocracking	5.8×10^5	6.4×10^6
Hydrogen generation	0.24×10^3 (million tons/h)	10.1×10^6
Kerosene desulfurization & treating	388.3×10^4	1.4×10^6
Cracking feed/vacuum gas oil desulfurization	43.4×10^5	1.3×10^6
Catalytic reforming	1×10^4	0.6×10^6
Gasoline/naphtha desulfurization & treating	23.1×10^4	0.2×10^6
Sulfur recovery unit	–	0.1×10^6
Tail gas recovery unit	–	0.8×10^6
Others	2.7×10^2 (million tons/h)	7.9×10^6

The atmospheric distillation in SHU processes some 100 Mbpsd less than MAA, which explains the difference in the fired heaters duty between them (Table 3 and Table 5). However, SHU process about three times more feed in its vacuum distillation units which are quite an energy intensive process. In addition, other processes in SHU lacking in MAA add to the total load of its fired heaters duty, i.e. Vacuum gas oil desulfurization. Fig. 6 and Fig. 7 show the CO_2 emission from SHU refinery fired heaters. The bulk of emissions correspond to the HP units, distillation columns and hydrocracking, similar to the case of MAA. Table 6 shows the fired heaters duty for MAB with respect to unit operation section and Fig. 8 and Fig. 9 show the CO_2 emission from heaters in MAB and percentile distribution according to units' category. It was witnessed that all refineries showed major heaters contribution from hydrogen generation units, topping and hydrocracking. Coking units in MAB present themselves as a major source of carbon emission from heaters and boilers contributing to 10% of the carbon emissions load. HP and topping units contribute in a 24% and 28%, respectively (Fig. 9).

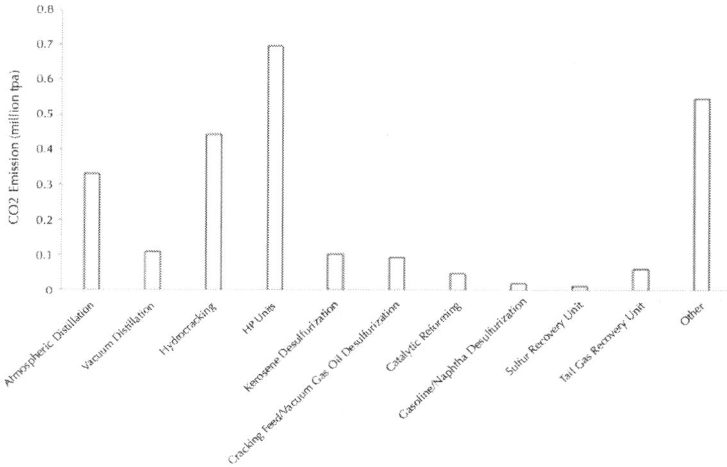

Figure 6: CO_2 emission from heaters and boilers in SHU refinery categorized per combined units section.

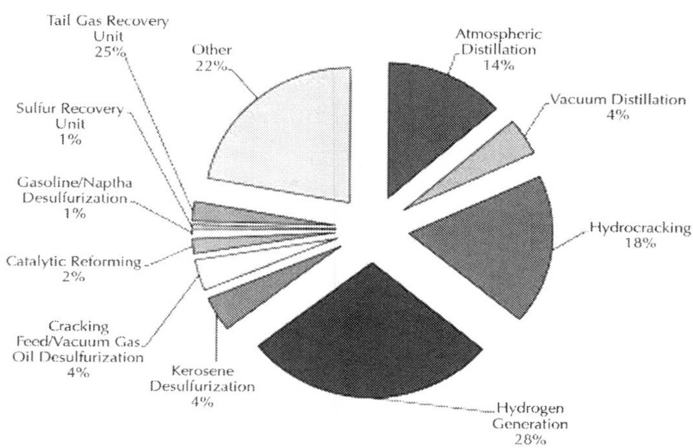

Figure 7: Percentile distribution of CO_2 emission from heaters and boilers in SHU.

Table 6: Combined fired heaters duty in MAB categorized based on process type

Process unit type	Throughput (bpsd)	Fired heaters duty (GJ/year)
Atmospheric crude distillation	3.3×10^5	5.6×10^6
Vacuum distillation	1.5×10^5	3×10^6
Hydrocracking	1.6×10^5	2.2×10^6
Hydrogen generation	69 (klb/h)	7×10^6
Coking	7.8×10^4	3.1×10^6
Residual desulfurization	2.5×10^5	2.1×10^6
Gasoline/naphtha desulfurization & treating	8.9×10^3	1.6×10^5
Distillate desulfurization & treating	3.9×10^4	2×10^5
Kerosene desulfurization & treating	4×10^4	2.6×10^5
Others	6.8×10^2 (klb/h)	6.8×10^6

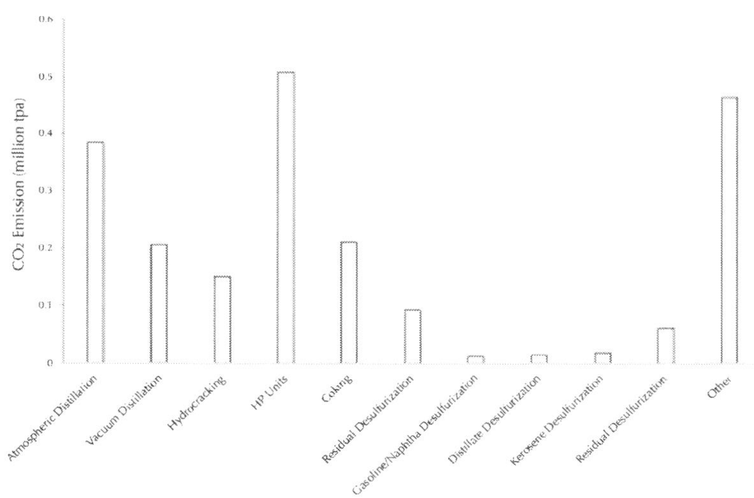

Figure 8: CO_2 emission from heaters and boilers in MAB refinery categorized per combined units section.

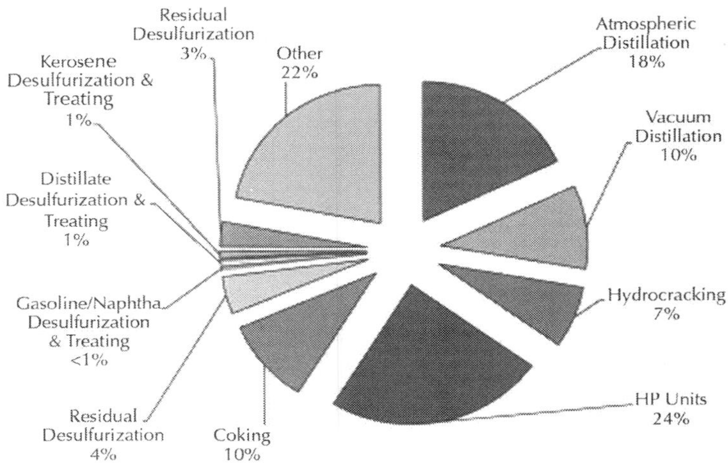

Figure 9: Percentile distribution of CO_2 emission from heaters and boilers in MAB.

Process heaters contribute to a large proportion of CO_2 emission from refineries as previously stated by a number of authors (Stockle and Bullen, 2008, van Straelen et al., 2009 and Johansson et al., 2012). However, individually each process heater emits a small quantity of CO_2. This makes pre-combustion capture for carbon in process heaters economical viable solution to produce a concentrated stream of CO_2gas for future utilization. Alternatively, a different fuel could be used in the heaters especially considering the fraction of associated gas in Kuwait's fields or considering hydrogen fuel. Lowe et al. (2011) modeled the performance of refinery heaters using hydrogen fuel (95% H_2 and 5% C_1, H_2O and N_2) instead of typical refinery fuel gas. Heating efficiency, on a lower heating value basis, slightly increased when switching from refinery fuel gas to hydrogen fuel. In addition, NO_x and GHGs decreased in emission which gives way for carbon emission mitigation in refineries. This might be appropriate in Kuwait if H_2S reformers are to be considered in the future plans of the downstream industry, decreasing CO_2 emissions from heaters and having a higher H_2 supply in the refinery as a utility gas.

Flaring Activities from Existing Refineries in Kuwait

Flaring in refineries and other industrial facilities occur as an alternative to venting hydrocarbons to the atmosphere. It is an important safety measure for controlling operation shutdowns and start-ups, in addition to controlling operating pressures. Flaring results in the combustion of hydrocarbons with CO_2 and steam as products. A survey was conducted in this work to determine the main reasons for flaring to occur in refineries. These could be summarized as fluctuation in H_2 header pressure in ARDS, makeup gas compressor leaks in hydrocrackers, HP units compressor discharge and maintenance of topping and other units. Martin et al. (2014) reported a correlation based on emission inventory analysis that estimates the amount of greenhouse gases (GHG) emitted by refineries flares. This was used in this study to estimate the amount of CO_2 emitted based on the following:

$$\mathrm{FCE} = 3.12\,RT \tag{5}$$

where FCE is the flares CO_2 emissions (kg) and RT is the refinery's throughput in tons. Kuwait's crude (more commonly termed Kuwait Export Crude, KEC) is of a 30.2 API in recent years with an estimated density of 875 kg/m^3. Hence, each refineries throughput in tons per day (tpd) is estimated and the amount of CO_2 from flaring is calculated based on reported data (2012). The estimated specific CO_2emission from flaring in MAA, SHU and MAB were estimated to be (ktpa): 72.8, 31.6 and 42.7, respectively.

Acid Gas Removal Carbon Emissions

Acid gas removal (AGR) processes using various types of amine solvents are common in petroleum refineries and are a crucial part of the operations in gas treatment plants. Natural gas consists of hydrocarbons and acid gases (hydrogen sulfide (H_2S) and carbon dioxide (CO_2) and the process of removing hydrogen sulfide from it, is commonly referred to as sweetening processes. In MAA, propane, butane and gasoline are cryogenically extracted through extraction/dehydration, cooling/chilling, fractionation and product treating. Moisture and liquids are removed from the entering gas to prevent freezing in the cold section of the plant. Diethanolamine (DEA) solution is used in this unit which is of a 230 MMSCFD capacity. The gas feed enters the absorber unit where H_2S and CO_2 are absorbed using DEA. The treated gas is fed to a separator to separate the carried DEA. Using the absorber balance equation for CO_2, one can estimate the amount of CO_2 exiting the system as follows (CFR, 2014):

$$E(CO_2) = V_{in}\left[\frac{vol_{in} - vol_{out}}{1 - vol_{out}}\right] \tag{6}$$

where $E(CO_2)$ is the amount of CO_2 emitted from the system (SCFD), V_{in} is the flow rate of the gas in the unit (SCFD), vol_{in} is the mol fraction of CO_2 in the entering stream and vol_{out} is the existing mol fraction taken as (1.2% maximum). Estimated amounts of CO_2 are dependent on the CO_2 content in the treated gas which ranges from 7.19 to 9.28 mol%. The amount of CO_2 vented to the atmosphere was estimated to be 0.15–0.32 million tpa depending on the treated gas properties.

MAA refinery also treats acid condensate in a similar process. A density of 750 kg/m^3 was assumed for the condensate (IPIECA, 2014). From the unit's inert balance, the CO_2 amount emitted was estimated to be between 0.24 and 0.83 million tpa depending on the condensate properties and composition. As for SHU refinery, a similar methodology was used to estimate the CO_2 vented due to acid gas processing. The gas and condensate flow rates are reported as 48.5 MMSCFD and 4 Mbpd, respectively. A CO_2 content in the acid gas and condensate were reported as 11 mol% and 5.9 mol%, respectively (Al-Qattan, 2014). The amount of CO_2 out from gas and condensate treated in SHU was estimated to be 1.07×10^5 and 1.02×10^5 tpa, respectively.

Load of Electrical Import to the Refineries

Refineries typically produce power in the form of electricity, hence the confusion typically occurs with utilities as produced or consumed. In Kuwait, refineries do not produce electricity; in fact they purchase electrical supply from the national grid of the Ministry of Electricity and Water (MEW). According to the Climate Registry Default Emission Factors (CRD, 2012), Kuwait's emission factor for electricity production is 870 g CO_2/kWh based on fossil fuel burning. Using the CRD (2012) emission factors the amount of CO_2emitted due to the electricity consumption in MAA, SHU and MAB are 771, 379 and 574 mtpa, respectively.

ANALYSIS AND DISCUSSION

Table 7 summarizes the results obtained for the three refineries (million tpa). The total CO_2 emission was considered without the electricity input by the MEW. This load contributes to the total carbon footprint of the refinery, but cannot be considered as a source of emission by petroleum processing within the studied boundary, i.e. the refinery processing scheme. Typically it is considered as an indirect burden (HCPC, 2010). The total amount of CO_2 emitted from MAA was estimated to be 3.78 million tpa. As previously indicated; utilities pose the major source of CO_2 in refineries. Fig. 10 shows the contribution of each source to the total amount of CO_2 emitted in each refinery. As indicated, the type of refinery influences the distribution of the total pool of CO_2 emissions. MAA, includes an FCC unit which contributes to 6% of the total CO_2 emissions. MAA's CO_2 source distribution is inline with past published results. This is due to the large number of CO_2 sources it hosts (Stockle and Bullen, 2008, DNV, 2008, van Straelen et al., 2009, van Straelen et al., 2010 and Johansson et al., 2012).

Table 7: Carbon dioxide (CO_2) emission (million tons per annum) from the three existing refineries in Kuwait

Source	MAA	SHU	MAB
Fluid catalytic cracking (FCC)	0.33	N/A	N/A
Fired heaters	3.55[a]	2.45	2.12
Acid gas removal (AGR)b	0.03–1.16	0.01	N/A
Flaring	0.07	0.03	0.04
Hydrogen production (HP) units	0.71	0.73	0.72
Electricity Load	770	379	573
Total (million tons per year)[c]	3.78	3.22	2.88
% of Total in Kuwait (w/o Electrical Import)	48	28	24
Specific emission rate (ton CO_2/bpd)	8.1	16	1.6

[a]Including gas plant side (acid gas sweetening processes and condensate processing).

[b]Total of acid gas removal and condensate sweetening processes combined.

[c]Based on maximum CO_2 content in acid gas without the electricity load.

There is a lack in open literature regarding utilities contribution to CO_2 emission and which heaters and boilers typically emit the largest amounts of CO_2 with respect to unit type and capacity. However, it is understood that unit operations in refineries are quite energy intensive. Heaters and furnaces (unit utilities) represent 62–74% of CO_2 in the refineries of Kuwait. Past research also indicated that heaters contribute to anything between 26% and 65% of the CO_2 emissions from a petroleum refinery (DNV, 2008, van Straelen et al., 2009 and HCPC, 2010). Refineries in Kuwait contribute to slightly larger degree when it comes to process heaters, which can be attributed to two reasons. Firstly, the capacity of Kuwait's refineries is considered on the larger side worldwide; and secondly, both SHU and MAB are energy intensive refineries dealing with heavier crude oils with low APIs and high unit utilization capacities (>70%).

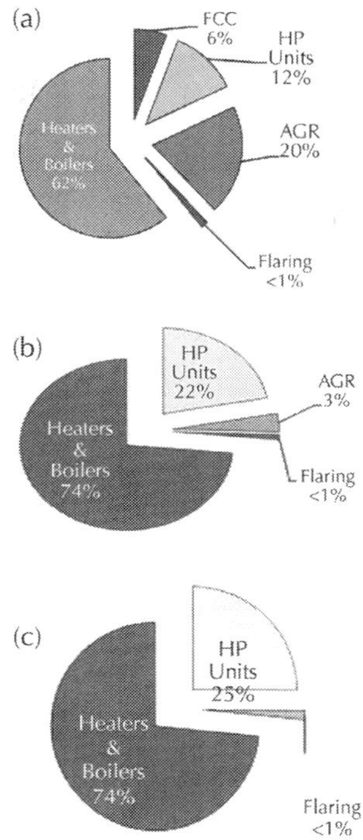

Figure 10: Contribution to CO_2 categorized by source in (a) MAA, (b) SHU and (c) MAB.

Furthermore, the requirement of lower sulfur content in products is now increasing which requires certain unit utilities to be supplied to meet such demands. This is more commonly termed as the '*petroleum refining paradox*' (Johansson et al., 2012). Past assessment of EU refineries given by Johansson et al. (2012) included mostly refineries with no vacuum distillation, hydrotreating units or FCC units; which is the case of most European petroleum refineries. Only 18 refineries (out of 144 in the EU) operate complex systems that include such units. Hence, source distribution when considering complex configuration refineries (as in the case of Kuwait) shows a larger contribution of heaters to the total emission load. This is evident in the case of MAA where an FCC unit is in operation, in addition to hydrocrackers and vacuum distillation units.Johansson et al. (2013) reported

CO_2 emission for two refineries in the EU. The first was a hydro-skimming refinery (which includes in addition to crude distillation, a catalytic reforming, hydrotreating units, naphtha upgrading units and product blending capabilities are included in its scheme) with a capacity of 6 million tpa and a CO_2 emission rate of 0.5 thousand tpa. The second is a complex refinery of a capacity equal to 11.4 mtpa and an emission rate of 1.9 mtpa. By comparison to Kuwait, where availability of feedstock is an influential factor, emission rates of CO_2 are minimal. This needs to be considered by the energy policy makers in the country to implement a more green operation in the downstream sector. When considering the basis of estimating the heaters emission, fuel oil was considered as the fuel of choice for heater operation. A noticeable reduction (≈25%) is noticed in CO_2 emissions when natural gas emission factor was used (Fig. 11). This points toward more carbon savings in terms of developing new energy policies in the country. The majority of associated gas in the country is not utilized and flared in gathering centers. This could be utilized to fuel some heaters in the downstream industry to minimize the carbon load.

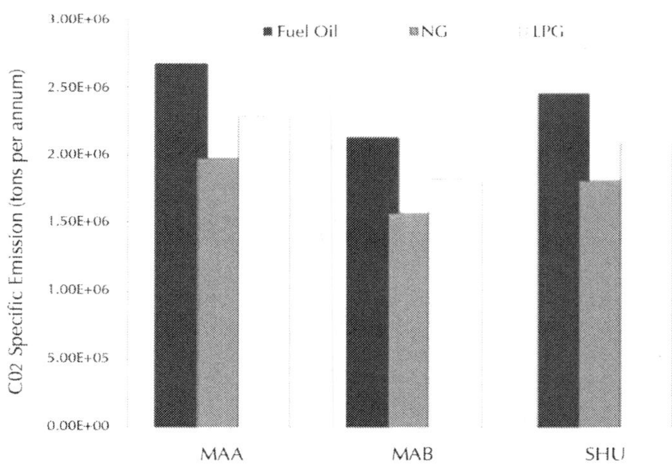

Figure 11: Variation in using fuel oil, natural gas (NG) and liquefied petroleum gas (LPG) as a fuel for process heaters in refineries.

Flaring in Kuwaiti refineries is highly sensitive to daily operations. As expected, since the amount of CO_2 emissions is directly related to the processing capacity of a refinery, MAA has the major share of such emissions (≈50%) due to its large capacity. Flares in refineries are high in concentrations of hydrocarbons (typically rich in methane, C_1), which with complete combustion with methane can produce in a 1:1 ratio carbon dioxide (CO_2) and with ethane in a 1:2 ratio (Fahim et al., 2010). Incomplete combustion can also occur if insufficient air is supplied. A flare system has multiple flares to treat the waste gases in a refinery. CO_2 can range between 0.02% and 2.8% depending on the type of operation, hydrocarbons content and operating conditions (Cheremisinoff and Rosenfeld, 2009). Fisher and Brennan (2005) discussed the benefits of installing flare gas recovery units (FGRUs) in refineries. Such systems capture gases (pre-flaring) for use in the refinery or for cleanup and polishing. Furthermore, flare gas may encompass substantial heating value and could be used as a fuel within the refinery to reduce the amount of purchased fuel. This strategy would benefit the refineries under consideration in this work, since flue gas will be reduced and carbon emissions will reduce in the global CO_2 estimation. Peterson et al. (2007) noted that the gas processing capacity of the FGRU adjusts to maintain a positive pressure on the flare header upstream from the existing liquid seal drum. This positive pressure is a safeguard that ensures that air will not be drawn into either the flare system or the FGRU.

HP units' contribution in the downstream global emissions is evenly spread between the three refineries (33%). Considering the capacities of the three refineries in Kuwait, the intensity of the operations in both SHU and MAB, with the addition of the impurities and level of contaminants in the processes gas in the reformers could explain the high percentages these refineries show (Table 7). However, due to the larger number of units (including the gas plant side), MAA utilities contribution is the largest amongst studied refineries when compared to the total carbon emissions of the downstream industry in the country (50%). For a 250 Mbpd capacity refinery, HCPC (2010) indicated that HP units can contribute by 3.3% of the total refinery carbon load or between 2.6% and 6% when scaled to Kuwait's refinery capacities. Other researchers have indicated that HP units could contribute to anything between 5% and 20% (Johansson et al., 2012). In Kuwait, 12–25% of the refineries CO_2 emissions were due to HP units. The estimate is slightly higher in the case of MAB and SHU. This is primarily due to the varied number of sources in MAA of CO_2. AGR activities and flaring are highly dependent on the processing capacity of each process block in the refinery. However, the purity of the feedstock is crucial in AGR. This can influence regeneration of the solvents (in stripping columns), hence influence the CO_2 emitted.

Heaters and boilers in refineries are crucial in operation and it is paramount to consider their contribution to the carbon load. However, to be able to make a sound judgment based on the operation of any given unit, we need to estimate the amount of CO_2 emissions based on the whole operation, i.e. utility, process emissions, impact from support processes, etc. Based on the work conducted in this study, emission factors for major units were developed to be able to estimate the CO_2 emission. Table 8 summarizes the CO_2 emission formulations based on the work conducted in this study. Distillation units were formulated based on the heaters and furnaces direct heat (energy) supply to the units (Table 8). Atmospheric and vacuum distillation use about 45% of refineries energy due to topping separation units (API, 2000). In Kuwait, refineries distillation units consume 18–28% of the refineries energy. Steam methane reforming (SMR) is also one of the most energy intensive operations in refineries, where hydrogen (H_2) gas is produced. Table 8 details the contribution of units' utilities and the contribution of hydrogen production, in terms of specific CO_2 emission. Hence, it includes all contributions from the unit itself and support units to the carbon emissions load. Refineries face a lot of challenges in carbon emissions mitigation especially when considering the changes in fuel mix, energy process, increasing fuel quality demands and heavier crude feeds. Moreover, environmental quality specifications for diesel and gasoline affect other unrelated ones, such as lubricity and gasoline octane number (Szklo and Schaeffer, 2009). For example, FCC gasoline contributes to the sulfur content by 85–95% (Bruent et al., 2005). However, due to its high olefin content (20–40 wt%), gasoline has a high octane number as a standalone product. Hence, to produce a low sulfur product, refineries need to consider optimizing the high energy requirement to its operation.

Table 8: Carbon dioxide (CO_2) specific emissions formulations developed in this work based on each major unit utilities and support units contribution

Process	Specific CO_2 emission[a]	Notes
Atmospheric distillation	CE = 1.4 TP	–
Vacuum distillation	CE = 1.5 TP	–
Hydrogen production (HP) via SMR	CE = 112 TP + 0.25 (HP)	Throughput and production rate are based on million tons per hour.
Hydrocracking	CE = 1.1 TP	Hydrocracking uses about 54 m^3 H_2/m^3 feed (Gary and Handwerk, 1994); which should be added to the formulation depending on the unit throughput.
Residual desulfurization	CE = 0.28 TP	Desulfurization H_2 flow should be added depending on the unit throughput.

[a]Where CE is the carbon emission (tpa), TP is the unit throughput (bpsd) and HP is hydrogen production in million tons per hour.

A number of influencing factors affect the specific emission rate of a refinery, especially the quality of the crude oil processed. When considering the type and configuration of a refinery and the properties of crude oil it processes, one can observe the influence of operational conditions on the emission rate. Crude feed density, sulfur content, light liquids ratio to other products and unit utilization capacity are all factors that influence CO_2 emissions. A correlation was established between these factors and the emission rates of US refineries by Karras (2011), with an increasing trend. In addition, lower API which indicates heavier crude and higher sulfur content influences the carbon emission of the refinery. Both these factors indicate a more intense energy requirements (energy per barrel) and severe process intensity (HCPC, 2010). Such an effect could be attributed to two major factors. Firstly, lighter, sweeter crudes require less conversion and desulfurization. Secondly, for lighter and sweeter crudes, the refinery's energy requirements are met by more low-carbon fuel gas and less coke, fuel oil and other higher-carbon streams. Other crude oil properties also influence the carbon emission of a refinery, consequently the specific emission rate, such as the distribution of hydrocarbons (i.e., naphtha, distillates, gas oil), and the type of heteroatom compounds.

A useful tool for refiners is to estimate the specific emission rate of a certain process (HCPC, 2010 and Pagano et al., 2010). This was achieved in this work by determining the emission factors for the most influential processes on carbon emissions (Table 8). Higher requirements for energy and utilities are going to be needed to achieve low sulfur in products. Current refineries in Kuwait have an estimated specific emission rate ranging between 1.6 and 16 ton of CO_2/bbl processed (Table 7). The specific emissions of a refinery can range from 0.02 to 0.82 ton of CO_2 per ton of crude oil processed (Ecofys, 2009). Hence, if a density of 875 kg/m^3 was assumed for each barrel of crude and a volume of 0.16 m^3/bbl, the specific emission rate could be allocated to the range between 0.1 and 4.4 ton of CO_2/bbl crude processed. Pagano et al. (2010) reported a specific emission rate of US refineries operating with API of 32–38.3 with residue gasification to be 0.7 ton of CO_2/ton of crude oil processed (3.82 ton of CO_2/bbl crude processed). Lu et al. (2007) reported an average emission rate of the Illinois basin refineries in the US as 11.4 ton of CO_2/bbl processed of heavy crude oil. Hence, a strong relationship between API and emission rates of refineries can be witnessed in past work and the estimates of Kuwait are on the same trend, where higher API of crude oil corresponds to a lower emission rate. This can be clearly witnessed in the case of the state of Kuwait where estimated API of processed crude could fall below 22 and averages around 30 (Al-Fadhlee, 2014).

CHALLENGES IN CO_2 MITIGATION IN KUWAIT AND APPROPRIATE ROUTES OF CAPTURE

The main challenge in reducing CO_2 emission from refineries is twofold; the optimization of refineries utilities and energy consumption; and capturing CO_2 of similar quality to be mixed in a single stream. Better energy management in a refinery leads to a more cost effective operation and reduction in the total carbon footprint of the different processes. Heaters pose a major emission source in Kuwait's refineries. Hurst and Walker (2005) suggested that CO_2 should be collected from all the stack gases from heaters across the refinery and duct them to a central location. CO_2 is then separated and compressed and gases are moved across the site with the aid of fans, moving the stack gases to the central treatment process. The detailed design for this work seems to be unfeasible at larger scale refineries as indicated by van Straelen et al. (2010), since the space for large diameter ducting will be a challenge. Hence, the focus should be only on large emission stacks. DNV (2008) suggested that this lack of agreement is an indication that such problems are site specific which warrants further investigation. In addition, the quality of the captured stream is paramount. Heaters have a typical concentration of CO_2 in the range of 8–10%. The requirement to remove nitrous and sulfur oxides (NO*x* and SO*x*) from the stack gases before the CO_2 absorber to reduce degradation of the absorption solvent is another irksome task to manage. With the vast array of refineries configurations with numerous emission points of different sizes, the capture rate at refineries will be limited and determined by the quality of *hard to connect* emissions points rather than the capture efficiency of the capture technology/process (DNV, 2008). Oxy fuel combustion offers another solution for CO_2 capture from heater and boilers. Wilkinson et al. (2003) and Allam et al. (2005) proposed that all heaters and boiler on site are modified for firing with pure oxygen, produced at a central location. Flue gases from combustion will be treated at locations near the stacks (for water removal) and CO_2 will be compressed to 30 bars. The production of the oxygen will require considerable amounts of additional (electrical) energy. Hence, an increase in CO_2 emissions is anticipated.

The emerging developments in the past few years have focused on developing technologies for flue gas CO_2 reduction among other carbon emission sources. This is mainly the case for FCC, where this process contributes (up to) 20% of the total CO_2 load from a refinery. The amount of coke and subsequent CO_2produced in the reactor is a direct result of the operating conditions. The unit yields are directly related to the quality of the feed and the operating severity set by the enthalpy balance. Wolschlag et al. (2009) have discussed the opportunities of carbon emission reduction in FCC and allocated such techniques to the following: Firstly, improved energy optimization typically achieved via 'Gap analysis' is suggested to identify root causes (insufficient maintenance, inefficient energy system design, inefficient process operation, procedures and control or outdated technology). Secondly, CO_2 reduction through energy efficiency revamps. This is mainly concerned with power recovery in FCC units. An example of such is the five-body train of a hot-gas expander, main air blower, start-up steam turbine, motor/generator and gear box as necessary. In this configuration, the expander is coupled to the main air blower and provides a direct transfer of energy to the shaft to minimize power transfer losses to the air blower.

Legislations on refineries and environmental rules and regulations have given rise to CO_2 reduction planning in the past decade or so. This will drive operational changes to improve process efficiencies, resulting in simulation packages development for carbon reduction schemes. This is especially true nowadays where heavier crudes are considered to be the sole feedstock in many refineries (new or revamped). The use of a simulation package has led to 50% reduction in flaring after implementing an online program (Moore, 2004). The simulation package includes inferential fuel gas quality and volume imbalance prediction, and a furnace-monitoring and advisory system. Additional benefits reported include: improved furnace efficiencies, more consistent product qualities (for example, where the process furnace is a feed heater or re-boiler) and more reliable process operation. Coupled H_2 and CO_2 management has also been studied recently in refineries. By developing a mixed integer non-linear program (MINLP), Al-Hajri (2008) has optimized refinery operations by controlling fuel, FCC and hydrogen reduction units. This optimization strategy could benefit the refineries in Kuwait immensely to manage their carbon footprints. Energy management in an oil refinery is an integral part of CO_2 reduction strategies, especially with new regulations governing sulfur content of gasoline and diesel in the past few years. This leads to increase in capacity. It is estimated that the total amount of crude oil processed in US refineries increased by comparison to the start of 1990s by 5%, complying with gasoline and diesel specifications. In Brazil, an increase of 1.5 million tons of CO_2 was estimated due to change in gasoline and diesel specifications in 2009. This leads directly to more energy intensive operations that require more hydrogen for cleaner fuel production. It is estimated that a refinery uses 7–15% of its crude oil input in its processes in direct or indirect heat requirements. To deal with energy management challenges, technologies that show potential can be stated as follows:

- Alternatives for saving energy at refineries
- Less severe alternatives for processes intensive operations (e.g. hydrotreating)
- Oil gasification and CO_2 removal at the refinery

Szklo and Schaeffer (2009) further described the first two options as achievable in the near to medium future, whilst the last could be said that it is developmental stages. Petrick and Pellegrino (1999) stated that a 15–20% target reduction of energy use is achievable from US refining activities, with emphasis on waste heat recovery utilization in petroleum refineries. Improvement of heat integration and waste heat recovery has been discussed in recent literature for more energy efficient systems in refineries (Szklo and Schaeffer, 2009). This could be allocated to the following treatment options:

- Use of waste heat in absorption refrigeration systems and utilize heat to pre-heat feeds.
- Improvement of furnace efficiencies combined with computer controlled combustion.
- Direct feed of intermediary products without storage to recover loss heat (omitting cooling stage).

Modern refineries have higher emissions due to heavier crudes use and the production of light products, especially gasoline and diesel. Catalytic reformers, FFC units and vacuum distillation units could also be equipped with combined heat and power (CHP) units with CO_2 capture schemes (IEA, 2004). These units represent 30–40% of the refinery energy consumption. This nominates them to be in the top of the list of unit operations that require energy management and carbon mitigation strategy.

CONCLUSIONS

Kuwait is an oil dependent country that hosts mega refining networks which operate three of the world's largest refineries, namely Mina Al-Ahmadie (MAA), Shuiba (SHU) and Mina Abdullah (MAB). These refineries have been studied and a critical analysis of their current CO_2 emission sources has been discussed in this work. The emission rates of these refineries were estimated at 3.78, 3.2 and 2.88 million tpa. The specific refinery emission rate could be estimated for MAA, SHU and MAB at 8.1, 16 and 1.6 ton of CO_2/bbl processed per day. These estimates show that Kuwait's downstream sector is a major carbon emitter that needs energy and operation optimization in a more environmentally friendly manner. The analysis revealed that utilities (mainly fired heaters) in current operating refineries constitute the major share of carbon emissions (62–74%). This could be managed with an energy optimization strategy and a collection of stack gases that could reduce the carbon footprint of this structure in the near future. HP units, which can contribute up to 25% of current refineries carbon load, can be an ideal candidate for capture projects in the future. Operational utilities and space availability are two major advantages for such units to be considered for future capture projects. Optimally, carbon emissions will reduce in Kuwait after taking into account direct heat requirements of units in the near future for better utilization of recovered heat. This will pave the way for future processing of crude in the country, especially when considering the lower API feedstock refineries are starting to process.

ACKNOWLEDGMENTS

The author wishes to thank the Kuwait Institute for Scientific Research for funding project PF066K. Eng. Ahmed Al-Terkait (Chem. Eng./Energy Management Eng.) is gratefully acknowledged for his contribution to this work for without which it would have been impossible to conduct. Eng. Ayed Al-Fadhlee (Kuwait Petroleum Corporation) has also graciously contributed to this workload and is thanked for his support. Eng. Faisal Al-Qattan (R&T Department-KNPC) is gratefully acknowledged for his contribution in providing process and operational manuals of MAA, SHU and MAB.

REFERENCES

1. Al-Ajmi, A.F., 2010. Clean fuel project: strategy and economics, Kuwait National Petroleum Company (KNPC). In: Proceedings of OAPEC/NOGA Joint Conference, Bahrain, 25–27 October.
2. Al-Fadhlee, 2014. Personal communication, Kuwait Petroleum Corporation, International Marketing Division.
3. Al-Hajri, I., (PhD thesis) 2008. Integration of Hydrogen and CO2 Management within Refinery Planning. Chemical Engineering Department, University of Waterloo, Canada, pp. 2008.
4. Allam, R., White, V., Stein, V., McDonald, C., Ivens, N., Simmonds, M., 2005. Revamping heaters and boilers to

oxyfiring-producing oxygen by the ITM technology. In: Rubin, E.S., Keith, D.W., Gilboy, G.F. (Eds.), Carbon Dioxide Capture for Storage in Deep Geological Formation, vol. 1. Elsevier, Oxford, UK, pp. 513–535.

5. Al-Qattan, F., 2014. Personal communication, KNPC, R&T Department – Process Manuals of Refineries.
6. Al-Salem, S.M., Khan, A.R., 2010. Monitoring and modeling the trends of primary and secondary air pollution precursors: the case of the state of Kuwait. Int. J. Chem. Eng., ID# 879836.
7. API – American Petroleum Institute, 2000. Technological Roadmap, Houston, TX, USA.
8. Bruent, S., Mey, D., Perot, G., Bouchy, C., Diehl, F., 2005. On the hydrodesulfurization of FCC gasoline: a review. Appl. Catal. A: Gen. 278, 143–172.
9. CA, 2008. Climate Act (CA) for the United Kingdom (UK). Policy: Reducing the UK's Greenhouse Gas Emissions by 80% by 2050, Available from: https://www.gov.uk/government/policies/ reducing-the-uk-s-greenhouse-gas-emissions-by-80-by-2050
10. CFR, 2014. Calculating GHG Emissions. Title 40. Cornell University, Available from: http://www.law.cornell.edu/cfr/text/40/98.233
11. Chehadeh, D., Kam, E., Mirza, Y., Al-Shumari, B., 2004. Experience in air emission inventory development for refineries of the Kuwait National Petroleum Company. In: Proceedings of Energy Conservational and Environmental Protection in the Petroleum Industry Seminar, Organization of Arab Exporting Countries (OAPEC), Cairo, 6–8 June.
12. Cheremisinoff, N.P., Rosenfeld, P., 2009. Handbook of Pollution Prevention and Cleaner Production – Best Practices in The Petroleum Industry. Guidelines for Cleaner Production., pp. 179–225.
13. CRD, 2012, January. Climate Registry Default Emission Factors. De Mello, L., Pimenta, R., Moure, G., Pravia, O.,

Gearhart, L., Milios, P., Melien, T., 2009. A technical and economical evaluation of CO2 capture from FCC units. In: Proceedings of Greenhouse Gas Technologies, Washington, DC, USA, 16–20 November. DNV – Det Norske Veritas, 2008. Global Technology Roadmap for CCS in Industry Sectoral Assessment: Refineries. Report No. 12P5TPP-9. EC – European Commission, 2014. EU

14. Emission Trading System (ETS) Low Carbon Fuel Standards. Ecofys, 2009, November. Methodology for the Free Allocation of Emission Allowances in the EU: ETS Post 2012. Sector Report for the Refinery Industry.
15. Fahim, M.A., Alsahhaf, T.A., Elkilani, A., 2010. Fundamentals of Petroleum Refining. Environmental Aspects in Refining. Elsevier, pp. 423–455
16. FDEP – Florida Department of Environmental Protection, 2014. Available from: www.dep.state.fl.us Fisher, P.W., Brennan, D., 2005. Minimize flaring with flare gas recovery. Hydrocarbon Process. 81, 83–85.
17. Gary, J.H., Handwerk, G.E., 1994. Petroleum Refining: Technology and Economics, 3rd ed. Dekker.
18. HCPC – Hydrocarbon Publishing Company, 2010. Refinery CO2 Managing Strategies, PA, USA.
19. Hurst, P., Walker, G., 2005. Post-combustion separation and capture baseline studies for the CCP industrial scenarios. Carbon Dioxide Capture for Storage in Deep Geological Formation, vol. 1., pp. 117–131.
20. IEA, 2004. Prospects for CO2 Capture and Storage. Energy Technology Analysis. International Energy Agency (IEA) and 50 Process Safety and Environmental Protection 9 5 (2 0 1 5) 38–50 Organization for Economic Co-operation and Development (OECD).
21. IPCC – International Panel on Climate Change, 2005. IPCC Special Report on Carbon Dioxide Capture and Storage.
22. IPIECA, 2014. Oil and Gas Industry Guidance on Voluntary Sustainability Reporting. Appendix D: Measurements Units

and Conversion Factors, Available from: http://www.ipieca.org/system/files/uploads/IPIECA Reporting Guidancen -Measurement units.pdf

23. JNOC – Japan National Oil Corporation, 1998, March. Investigation of CO2 Sources in Kuwait, Final Report.
24. Johansson, D., Franck, P., Bernstsson, T., 2013. CO2 capture in oil refineries: assessment of the capture avoidance costs associated with different heat supply options in a future energy market. Energy Convers. Manag. 66, 127–142.
25. Johansson, D., Rootzen, J., Bernstsson, T., Johnsson, F., 2012. Assessment of strategies for CO2 abetment in the European petroleum refining industry. Energy 42, 375–386.
26. Kam, E., Mirza, Y., Chehadeh, D., Ashkanani, A., Marouf, R., Al-Muhareb, E., 2005. KNPC Air Emissions Inventory Development, Air Module Implementation and Risk Assessment. Kuwait Institute for Scientific Research, Report No. 7629, Kuwait.
27. Karras, G., 2011. Oil Refinery CO2 Performance Measurement. Report Prepared for the Union of Concerned Scientists.
28. LG, 2011. Liberty Gases. CO2 Fact Sheet, Available from: http://www.libertygases.com/carbon-dioxide.html
29. Lowe, C., Brancaccio, N., Batten, D., Leungb, C., Waibe, D., 2011. Technology assessment of hydrogen firing of process heaters. Energy Procedia 4, 1058–1065.
30. Lu, Y., Garner, D., Korose, C., Chen, S., Rostam-Abadi, M., 2007. Inventory of Industrial Stationary CO2 Emissions in the Illinois Basin, Summary Report U.S. DOE Contract: DE-FC26-03NT41994.
31. Martin, J., Lumbreras, J., Rodríguez, M.E., 2014. Testing Flare Emission Factors for Flaring in Refineries, Available from: http://www.epa.gov/ttnchie1/conference/ei12/poster/martin.pdf
32. Moore, I., 2004. Reducing CO2 emissions. Pet. Technol. Q., 1–6. Pagano, M., Di Pasquale, S., Gragnani, A., Ratan, S.,

2010. Refinery H2 and CO2 management: key players in the immanent GHG compliance. In: Proceedings of Hydrocarbon Processing International Refining Conference, Rome, Italy, 21–23 June.

33. Peterson, J., Tuttle, N., Cooper, H., Baukal, C., 2007. Minimize facility flaring. Hydrocarbon Process., 111–115.
34. Petrick, M., Pellegrino, J., 1999. The Potential for Reducing Energy Consumption in the Refining Industry. Report No. ANL/ESD/TM-158. Argonne National Laboratory.
35. Romano, M.C., Anantharaman, R., Arasto, A., Ozcan, D.C., Ahn, H., Dijkstra, J.W., Carboe, M., Boavida, D., 2013. Application of advanced technologies for CO2 capture from industrial sources. Energy Procedia 37, 7176–7185.
36. Salman, M., Oskui, R., 2004. Selected Processes for Recovery of Crude Oils in Kuwait. Phase I: Preliminary Study, Kuwait Institute for Scientific Research, Final Report No. 7238, Kuwait.
37. Stockle, M., Bullen, T., 2008. Integrating refinery CO2 reduction strategies into your refinery. In: Proceedings of ERTC Sustainable Refining Conference, Brussels, Belgium, 31 March.
38. Szklo, A., Schaeffer, R., 2009. Fuel specification, energy consumption and CO2 emission in oil refineries. Energy 32, 1075–1092.
39. USEPA – United States Environment Protection Agency, 2010, October. Available and Emerging Technologies for Reducing Greenhouse Gas Emissions from the Petroleum Refining Industry. Office of Air and Radiation.
40. van Straelen, J., Geuzebroek, F., Goodchild, N., Protopapas, G., Mahoney, L., 2010. CO2 capture for refineries, a practical approach. Int. J. Greenhouse Gas Control 4, 316–320.
41. van Straelen, J., Geuzebroek, F., Goodchild, N., Protopapas, G., Mahoney, L., 2009. CO_2 capture for refineries: a practical approach. Energy Procedia 1, 179–185.

42. Wilkinson, M., Boden, J., Gilmartin, T., Ward, C., Cross, D., Allam, R., Ivens, N., 2003. CO2 capture from oil refinery process heaters through oxyfuel combustion. In: Gale, J., Kaya, Y. (Eds.), Greenhouse Gas Control Technologies. vol. 1, pp. 69–74.
43. Wolschlag, L.M., Couch, K.A., Zhu, F.X., Alves, J., 2009. UOP FCC Design Advancements to Reduce Energy Consumption and CO2 Emissions. UOP LLC.

Citations

CHAPTER 1

Alberto Alva-Argáez, Antonis C. Kokossis, Robin Smith, The design of water-using systems in petroleum refining using a water-pinch decomposition, Chemical Engineering Journal, Volume 128, Issue 1, 15 March 2007, Pages 33-46, ISSN 1385-8947, http://dx.doi.org/10.1016/j.cej.2006.10.001.

CHAPTER 2

Marcel Joly, Roger Rocha, Luiz Carlos F. Sousa, Marcia T. Takahashi, Pierre N. Mendonça, Leonardo A.M. Moraes, André D. Quelhas, The strategic importance of teaching Operations Research for achieving high performance in the petroleum refining business, Education for Chemical Engineers, Volume 10, January 2015, Pages 1-19, ISSN 1749-7728, http://dx.doi.org/10.1016/j.ece.2014.11.001.

CHAPTER 3

Sylvie Le Borgne, Rodolfo Quintero, Biotechnological processes for the refining of petroleum, Fuel Processing Technology, Volume 81, Issue 2, 15 May 2003, Pages 155-169, ISSN 0378-3820, http://dx.doi.org/10.1016/S0378-3820(03)00007-9.

CHAPTER 4

Jian Song, Yan Li, Chun-wei Gu, Li Zhang, Thermodynamic analysis and performance optimization of an ORC (Organic Rankine Cycle) system for multi-strand waste heat sources in petroleum refining industry, Energy, Volume 71, 15 July 2014, Pages 673-680, ISSN 0360-5442, http://dx.doi.org/10.1016/j.energy.2014.05.014.

CHAPTER 5

Ricardo Rivero, Application of the exergy concept in the petroleum refining and petrochemical industry, Energy Conversion and Management, Volume 43, Issues 9–12, June–August 2002, Pages 1199-1220, ISSN 0196-8904, http://dx.doi.org/10.1016/S0196-8904(02)00008-0.

CHAPTER 6

Xiaoyu Liu, Dingjiang Chen, Wenjun Zhang, Weizhong Qin, Wenji Zhou, Tong Qiu, Bing Zhu, An assessment of the energy-saving potential in China's petroleum refining industry from a technical perspective, Energy, Volume 59, 15 September 2013, Pages 38-49, ISSN 0360-5442, http://dx.doi.org/10.1016/j.energy.2013.07.049.

CHAPTER 7

S.M. Al-Salem, Carbon dioxide (CO2) estimation from Kuwait's petroleum refineries, Process Safety and Environmental Protection, Volume 95, May 2015, Pages 38-50, ISSN 0957-5820, http://dx.doi.org/10.1016/j.psep.2015.02.011.

Index